AGRICULTURAL RESEARCH AND EXTENSION SYSTEMS

About the Author

Ram Bahal, an M.Sc. (Agri. Extension) from the University of Kanpur and Ph.D. from the University of Illinois, U.S.A., is a Principal Scientist at the Division of Agricultural Extension, Indian Agricultural Research Institute, New Delhi. Earlier he was a faculty at the C. S. Azad University of Agriculture and Technology, Kanpur. He is also North Indian Editor of *Indian Journal of Extension Education*. Dr. Bahal has more than 100 research papers, book chapters and books to his credit.

AGRICULTURAL RESEARCH AND EXTENSION SYSTEMS

Worldwide Study of Human and Financial Resources

Ram Bahal

Principal Scientist
Division of Agricultural Extension,
Indian Agricultural Research Institute,
New Delhi

CONCEPT PUBLISHING COMPANY, NEW DELHI-110059

Cataloging in Publication Data—DK
Courtesy: D.K. Agencies (P) Ltd. <docinfo@dkagencies.com>

Bahal, Ram, 1950-
Agricultural research and extension systems : worldwide study of human and financial resources / Ram Bahal.
p. cm.
Includes bibliographical references.
Includes index.
ISBN 81-8069-104-7

1. Agriculture—Research. 2. Agricultural extension work—Research. 3. Agricultural extension work—Finance. 4. Agricultural extension work—Personnel management. I. Title.

DDC 630.715 21

ISBN 81-8069-104-7

First Published 2004

Published and Printed by
Ashok Kumar Mittal
Concept Publishing Company
A/15-16, Commercial Block, Mohan Garden
New Delhi-110059 (INDIA)

Phones : 25351460,25351794
Fax : 091-11-25357103
Email : publishing@conceptpub.com

Foreword

This book is an excellent contribution to scholarship in the fields of agricultural research and extension. The author investigated two major independent variables, namely the relative investment of human and financial resources by 114 governments in their respective national agricultural research and extension system. The research sought to detect whether the relative level of research and extension investment by these national governments had an impact on national agricultural development, as measured by four dependent variables. These dependent variables included: cereal yields, farmer income, agricultural income, and fertilizer utilization as a surrogate variable for technology utilization.

In assessing national human resource investment, the author examined nine variables, including the total number of agricultural extension workers and agricultural researchers in each country, the number of female extension workers, the average educational level of field extension personnel, the percentage of subject matter specialists in relation to the total number of extension personnel, the amount of arable land per extension worker, the amount of arable land per researcher, the average number of farmers per extension worker, and the average number of farmers per researcher. In terms of national financial resource investment, the author analyzed the total agricultural research and extension expenditure for each country, as well as the average research and extension expenditure per farmer in each country, all financial data were converted into deflated 1980 U.S. dollars. In addition, the database was categorized by geographic region and level of economic development among the responding countries.

The book is organized into five chapters, including the introduction, review of literature, methodology, analysis and findings, and the summary and conclusion. The study analyzed the impact of different levels of human and financial resource investment on each of four agricultural development indicators. The statistical tools used in this analysis included Chi-Square, correlation, multiple regression and multiplicative statistics. The study findings are written and presented in non-technical English, so that the results can be easily be understood by both researchers and extension practioners.

This study establishes a clear relationship between national investment in research and extension systems and the relative impact of these investments on selected agricultural development indicators. Therefore, this book should serve as a valuable tool and provide useful insights to planners and policy makers, research and extension directors, as well as scholars and practioners who are concerned with agricultural development in poor countries.

The data for this study were collected by the Food and Agricultural Organization (FAO) of the United Nations through their resident representatives in each participating country. This study was carried out under my supervision at the University of Illinois at Urbana-Champaign during the period of 1989-1991.

Date: 2-11-2002

BURTON E. SWANSON
Professor of Rural Development
Department of Agricultural and Consumer Economics
University of Illinois at Urbana-Champaign

PREFACE

This worldwide study of agricultural research and extension systems was the first of its kind ever completed. The study had four objectives: (a) to determine the number of agricultural research and extension personnel worldwide and then, by using seven specific human resource measures, to determine the relationship of these variables with four agricultural development indicators, including total cereal yield, farmer income, agricultural income, and total fertilizer consumption; (b) to determine the quality (educational level) of extension personnel in different countries and then to relate this measure to selected agricultural development indicators; (c) to determine the level of agricultural research and extension expenditure worldwide and then to determine its relationship with the selected agricultural development indicators; and (d) to determine which of these human and financial resource variables were most useful, collectively, in explaining differences among each indicator.

Information about the human and financial resources among national agricultural extension systems was collected through the use of mail questionnaires, during October 1988 through December 1989 under the auspices of the Food and Agricultural Organization of the United Nations from 114 countries. The measures of agricultural research investment were taken from the *ISNAR Agricultural Research Indicator Series*. Data used to compute the demographic and agricultural development indicators were taken from published secondary sources. These data were analyzed using the Statistical Package for Social Sciences (SPSS) for the microcomputer. Two sets of analyses were conducted; one set involved the worldwide data set, while the other used a sub set of data involving only developing countries. Some of the significant findings of the study follow:

1. Human and financial resources of agricultural research and extension system were directly related with cereal yield, farmer income, and fertilizer consumption collectively.

2. The relative number of extension workers and researchers in developing countries was found to be very low in contrast

with the high income countries of North America, Europe and Asia. The ratios of fewer farmers per researcher and per extension worker were significantly related with all four agricultural development indicators individually.

3. Extension workers' educational level was significantly related to all four indicators.

4. Higher levels of agricultural research and extension expenditure per farmer were significantly related to higher cereal yields, farmer income, and fertilizer consumption.

Based on the analysis of these data, the following recommendations were made:

1. The number and quality of agricultural extension workers and researchers in developing countries should be strengthened.
2. To increase agricultural productivity and development, agricultural research and extension expenditure per farmer should be increased in most developing countries.

This book will be very helpful to the researchers, planners, policy makers in planning and decision-making regarding the allocation of the human and financial resources to the agricultural extension organisations in future. This book is an attempt to answer many questions at global level which were never answered to mention few; human resources of Agricultural Extension, their education level, percentage of female extension workers, financial resources of extension, expenditure per farmer etc.

Ram Bahal

ACKNOWLEDGEMENTS

This study was made possible through the efforts of numerous individuals and institutions worldwide. The author wishes to acknowledge, with sincere appreciation, all individuals and institutions who assisted, guided, and cooperated in this worldwide study.

The author expresses sincere appreciation to Dr. B. E. Swanson who provided the author with an excellent opportunity to study at the University of Illinois and who served as academic advisor and chairman of the dissertation committee for this study.

Gratitude is also expressed to Dr. Edward W. Osborne, Dr. Klause G. Witz, Dr. Dale A. Law, and Dr. Sam Johnson III not only for their advice and guidance while serving on the dissertation committee, but also for their special efforts in promoting the quality of the study. The author expresses sincere appreciation to Dr. W. M. Maalouf and Dr. T. E. Contado at the Food and Agricultural Organizations of the United Nations for the substantial assistance they rendered during the research phase of this study. Thanks are also extended to Dr. John J. Nicholaides, Dr. Earl B. Russell, and Ms. Barbara Farner for their support and encouragement throughout this study. The author gratefully acknowledge the assistance of Ms. Lori Snipes in formatting the whole dissertation.

Also, the author would like to express his sincere thanks to all institutions that assisted in the successful completion of this study, including the Food and Agricultural Organizations of the United Nations; the International Service for National Agricultural Research; the University of Illinois; the Ministry of Welfare and Ministry of Agriculture of the Government of India, as well as the India Council of Agricultural Research and the Indian Agricultural Research Institute.

The author wishes to express his deepest appreciation to his family in India, especially his parents, wife Prem Pyari, son Rajesh Kumar and two daughters—Madhu Bala and Prem Bala—for their patience and encouragement during the study period.

Ram Bahal

CONTENTS

LIST OF TABLES

LIST OF FIGURES

1

INTRODUCTION

Informal agricultural extension services started in the early nineteenth century, but formal and organized extension services really began to be organized during the early and mid-twentieth century (Rolls, 1988, p. 3; Swanson 1984, p. 3; von Blanckenburg, 1984, p. 7; Kelsey and Hearne, 1963, p. 108) to improve the lives of rural people (Mosher, 1958 p.5) and more recently, to help meet the food shortage in many developing nations. Because of the varying nature of the socio-cultural, politico-economic and agricultural needs of different parts of the world, extension services started at different time periods. For example, in Europe agricultural extension work began in Ireland between 1847 and 1851 at the time of serious potato crop failure (Rolls et. al.). In France, Germany, North America and the Indian Sub-continent, various types of extension services started in the 1860s and 1870s. In the Caribbean by the late 1880s, in Egypt by the 1890s and in various African countries by the first decade of the 20th century. In fact, agricultural extension services started in all parts of the world by the beginning of the 20th century to enhance the agricultural production and improve the quality of life of the rural masses.

The human and financial resources available to agricultural extension are two key elements for the success of agricultural extension organizations in their endeavor. The quality and quantity of these resources has direct bearing on agricultural development. In other words, the success of agricultural development in general and agricultural extension in particular depends in large part upon the quality and quantity of human and financial resources available within the organization. Considerable human and financial resources have been used in agricultural extension organizations since then, but it was not specifically known the level of human resources were being used by the agricultural extension organizations until the International Directory of Extension organizations and extension training institutions of 80 countries was compiled by McCabe and Swanson in the beginning of 1975. Another study was conducted by Swanson and Rassi (1980), which provided information about national extension systems and human resources of 104 countries. But the current status of human and financial resources of agricultural extension and the relationship of these resources with the agricultural development process is still largely unknown. This study is, therefore, planned to study the current status of the human and financial resources of agricultural research and extension and their relationship to the agricultural development process.

Theoretical Rationale

The agricultural development process depends in large part upon the appropriate research and extension of the resulting technology to all categories of rural people for its adoption. Schultz (1956, pp. 54-78) noted that in agriculture, the extension services play an important role in transmitting new information and in developing new skills of farmers. Swanson (1984, p. 2), mentioned that "The past experience of agricultural extension demonstrates that the function of extension is essential to the agricultural development process". Contado (1990) reported that "there is evidence that changes in farm production and farm family income can represent extension impact" (p. 98). Jamison and Lau (1982) found that "extension was significantly positively related to efficiency" (p. 9). The World Development Report (1982) however, reported that "performance of extension services has been disappointing" (p. 72). Von Blanckenburg (1984) reported that "there have been many complaints that extension services, in view of their enormous task, are insufficiently staffed and that they lack adequate working facilities and finance to cover the recurrent cost" (p. 23).

Theoretical Rationale for Human Resources

Clark (1987) mentioned that "more and more organizations are realizing that their most important resources are human. This is specially true in an organization such as extension, whose major thrust is education" (p.25). The authors of a donor agency project document (cited in Ames, 1988) concluded that the real limitations on the institutional capacity for research and development were the result of insufficient resources; chief among these is the manpower limitation. According to Ames (1988, p. 3),

"In Africa today, human capital in agricultural sciences is in short supply. Trained administrators, researchers, and extension agents are not really available to formulate and direct agricultural policy nor are mid-level professional to implement policy. The fundamental human resource base supporting agricultural development is scarce".

Swaminathan (1985, p. 52) noted that "Women have also been generally bypassed by extension services. Their numbers in scientific and extension services are generally small". Swanson (1984, p. 6), mentioned that, "In most Third World countries, there was, and still is, a severe shortage of trained agricultural personnel". Brown (1983, p. 44) noted that "Third World countries have a critical need for extension subject-matter specialists who are located in the field and accessible to support community extension agents". Clark (1987, p. 25) noted that" Tight budgets, hiring freezes, and even layoffs are the reality for today's Extension Service. The question to be answered is: "How do we survive-much less flourish-in times of limited resources? The solution may be found in a careful study of human resources" (p. 25). These theoretical views were the impetus to study the human resources of agricultural research and extension organizations worldwide.

Theoretical Rationale for Financial Resources:

Financial allocation to agricultural extension is another resource on which the effectiveness of extension largely depends. von Blanckenburg (1984, p. 28) reported that "Government financial resources for rural development in poor countries are notoriously scarce". He further reported that such inadequacy is, for instance, striking in Sudan, Liberia, Mali, Morocco and many other countries. Hence, a number of programs seriously suffer from a shortage of funds.

Roberts (1989, p. 23) noted that often the ministries of finance and agriculture have managed to avoid committing a significant quantity of local funds to the system at its inception and have been unwilling or unable to finance the entirety at a later stage. The common result was that the staff of the extension projects continued to be employed, while the project's operating costs (maintenance and fuel for vehicles, staff allowances, and training) were cut. No system that relies on mobility and on constant interaction will be able to function under these conditions. To overcome this problem, he further mentioned that it is clearly necessary to confront the current financing requirements of a new extension approach squarely from the outset.

Evenson (1987) mentioned that "The ratio of research costs to extension costs is as much as 20 to 1 for the low industrialized countries and only 3 to 1 or so for the industrialized countries. Some of this difference is a quality difference (extension workers have quite advanced training in most industrialized countries and may have little training in low-income countries), and some is due to real cost differences. Many low-income countries do not have the capacity to train agricultural scientists and must incur high costs to train researchers and to purchase scientific equipment" (p.120).

Bembridge's study (cited in Bembridge 1989, p. 20) mentioned that "The annual budget of the extension services represents only a small percentage (approximately 20 %) of the total for agricultural development which in turn comprises a small percentage of the total national budget (± –8%). Available data suggests this is an extremely low level of intellectual investment and low commitment to agricultural development. These theoretical facts motivated to study the current financial resources of agricultural extension.`

Problem Statement

The human and financial resources of agricultural extension organizations are the backbone for the success of any extension program. There is general consensus in the literature that agricultural extension programs fail in many countries because of the inadequacy of the human and financial resources. The review of literature also indicated that there are very few empirical studies on this topic. The present human and financial resources of agricultural extension organizations as well as their relationship with agricultural development indicators is, therefore, not known to the national and international planners, policy makers, extension professionals, scientists, and donor agencies. In order to document the need and to allocate the human and financial resources to agricultural extension organizations, it is necessary for the above professionals to know the present human and financial resources situation of agricultural extension. It is equally important to know the relationships of these resources (human and financial) with selected agricultural development indicators and their impact on agricultural development. This is only possible if a worldwide study is conducted on this aspect. Therefore, the focal point or problem addressed by this research is to study the present human and financial resources of agricultural extension organizations worldwide, and to investigate the possible relationship or impact of human and financial resources on specific agricultural development indicators that reflect overall agricultural development of the country.

Purpose of the Study

The purpose of this study is three fold: The first is to investigate the present human resource situation of agricultural extension organizations worldwide. The second is to investigate the financial resource situation of agricultural extension organizations worldwide and, finally, the third is to assess the relationships between these resources (human and financial) and selected agricultural development indicators.

To achieve the above purposes, the following specific objectives will be pursued in this investigation

1. To estimate and compare the quantity (number) of existing human resources in national agricultural research and extension systems, categorised by geographic region and economic group and to determine their relationship with selected agricultural development indicators.
2. To estimate and compare the quality (educational level) of existing human resources in agricultural extension systems, categorised by geographic region and economic groups and to assess the relationship of these factors with selected agricultural development indicators.
3. To determine and compare the amount of financial resources being invested in agricultural extension organizations, categorised by geographic regions and research and economic groups and to assess their relationship with selected agricultural development indicators.
4. To assess the relationship between the agricultural development indicators and measures of the human and financial resources being invested in national agricultural research and extension systems.
5. Given past trends and different assumptions, to estimate the trend of financial resources required by agricultural extension organizations in future.

Significance of the Study

The findings of this study will provide first hand information to the national and international planners, policy-makers, administrators, researchers, extension professionals, and donor agencies about the following

1. The present human resource (quantity and quality) situation of agricultural extension organizations.
2. The present financial resources (amount) situation of agricultural extension organizations.
3. The relationship of these extension resources with selected agricultural development indicators, and comparative relational impact of these resources in conjunction with agricultural development indicators.

These findings will be helpful to planners and policy-makers in planning and decision making regarding the allocation of the human and financial resources to the agricultural extension organizations in future. It is expected that this study will be the first attempt to estimate the human and financial resources of agricultural extension.

Research Questions

The study will answer the following research questions

1. What is the quantity (number) of existing human resources of agricultural extension organizations in six geographic regions, five economic groups, and four different levels of agricultural population of the world and their relationship with agricultural development?
2. What is the quality (educational qualification) of existing human resources of agricultural extension organizations in six geographic regions, five economic groups, and four different levels of agricultural population of the world and its relationship with agricultural development?
3. How much finance is allocated to agricultural extension organizations in six geographic regions, five economic groups, and four different level of agricultural population of the world and its relationship with agricultural development?
4. Given past trends and specific assumptions, requirement trend of extension personnel by the agricultural extension organizations will be estimated.

5. Given past trends and different assumptions, requirement trend of finance will be estimated by the agricultural extension organizations.

Assumptions Underlying the Study

1. The 114 responding countries in this study are representative of the total population of countries with agricultural extension systems.
2. Responding countries provided reasonably accurate and complete data of human and financial resources requested.
3. Socio-cultural, politico-economic and agro-climatic conditions of responding countries are different from each other.

Limitations of the Study

Some responding countries may not be able to provide required data.

Definition of Terms

Agricultural Development

"Change in agricultural production techniques and in the structure of the agricultural enterprises towards a more desirable situation. Usually a situation in which farmers use more agricultural research findings and in which there is less subsistence and more market-oriented agriculture". (van den Ban & Hawkins, 1988)

Agricultural Extension

'An organized, non-formal educational activity, usually supported and/ or operated by government, to improve the productivity and welfare of rural people...' (Swanson and Claar, 1984, p. 15).

Arable Land

Refers to land under annual or temporary crops (double-cropped areas are counted only once), temporary meadows for mowing or pasture, land under market and kitchen gardens (including cultivation under glass), and land temporarily fallow or lying idle.

Different Levels of Agricultural Population

Is determined on the basis of percentage of labour force in agriculture and categorized in four categories as; low agricultural population countries = up to 20 per cent labour force, lower middle agricultural population countries=21-40 per cent labour force, higher middle agricultural population countries=41-60 per cent labour force, and higher agricultural population countries = more than 60 per cent labour force in agriculture in this study.

Economic Group

All member (159) and non member (13) countries of the United Nations are grouped according to their per capita income in five categories, i.e. low income, lower middle-income, upper middle-income, high-income, and industrial market economies in this study.

Economically Active Population in Agriculture

Is defined as all economically active persons engaged principally in agriculture, forestry, hunting or fishing.

Empirical Research

"Research that tries to develop generalizations by testing hypotheses with the aid of objective techniques, and in agreement with sensory perceptions". (van den Ban & Hawkins, 1988).

Extension

"A form of conscious social influence. The conscious communication of information to help people from sound opinions and make good decisions". (van den Ban & Hawkins, 1988)

Extension Organization

"The name given to the organization that directs itself mainly at planning, execution and evaluation of the activities of extension agents. This organization is also considered to be a (desirable) link between the developers and the users of new knowledge and insight". (van den Ban

Hawkins, 1988)

Fertilizer Consumption

Fertilizer consumption is measured as the number of kilograms of fertilizer used for each hectare of arable land.

Financial Resource

The recurrent funds allocated to and used by the agricultural extension organizations in paying salary, program costs, capital maintenance and other costs.

Geographic Regions

All member (159) and non member (13) countries of the United Nations are grouped in six geographic regions, i.e. Africa, Asia and the Pacific, Europe, Latin America and the Caribbean, North America, and the Near East in this study.

Human Resource

The administrative, professional, and technical extension personnel employed by agricultural extension organizations.

Independence Date

Refers to the date of political independence, unless radical change in the government, resulting in profound social change, has occurred since then.

Land Under Permanent Crops

Refers to land cultivated with crops that occupy the land for periods longer than one year and that need not be replanted after each harvest, such as cocoa, coffee, and rubber; also, it includes land under shrubs, fruit trees, nut trees and vines, but excludes land under trees grown for wood or timber.

Life Expectancy at Birth

Indicates the number of years a newborn infant would live if patterns of mortality prevailing for all people at the time of its birth were to stay the same throughout its life.

Organization

A formal group with an explicit goal, set of rules and procedures and a clear division of tasks with specific rights and duties.

Planning

A process of giving direction in all phases of carrying out a policy, from policy formation to policy execution.

Policy

A form of directed action which indicates as clearly as possible what one wants to achieve, how one wants to do it and how much time will be taken to achieve the set goals. It can be seen as an attempt at consciously influencing and controlling future situation.

Policy-making

The way in which a policy comes into existence. Observations can be made of the nature of the policy-makers, the division of power among those involved, the degree of integration of behaviour and attitudes and the degree to which the behaviour is directed at achievement of the objective.

Problem

The reasons why a person or organization is unable to change an existing situation to a desired situation.

Quality

The educational qualification and training acquired by extension personnel is referred to quality in this study.

Quantity

Total number of technical and professional level extension personnel involved in agricultural extension profession.

Subject Matter Specialist

The Subject Matter Specialist (SMS) provides technical training and guidance to extension workers, has an important role in the formulation of production recommendations, and is a focus of links between extension and research.

Total Agricultural Population

Is defined as all persons depending for their livelihood on agriculture. This comprises all persons actively engaged in agriculture and their non-working dependents.

Tractor Density

These data generally refer to total wheel and crawler tractors (excluding garden tractors) used in agriculture.

Village Extension Agent or Worker (VEW)

Generalist extension agent at the local level, which might include several villages, responsible for the direct contact with farmers.

2

REVIEW OF LITERATURE

Since its inception in the late 19th century history suggests that agricultural extension has contributed a great deal in agricultural and rural development (True, 1928; Rolls, 1988, p. 3; Swanson 1984, p. 3; von Blanckenburg, 1984, p. 7; Kelsey and Hearne, 1963, p. 108). In order to make more efficient and effective agricultural extension systems in future, it is necessary to examine the adequacy of two of its major components i.e. human and financial resources.

Human resources of agricultural extension, which include such factors as : (a) the relative numbers of research and extension personnel, (b) the gender breakdown of research and extension personnel, (c) the quality or educational qualifications of extension personnel, (d) the amount of work carried out by extension personnel, and (e) incentives for extension personnel.

The financial resources of agricultural extension includes (a) the relative level of financial resources allotted to research and extension, (b) the adequacy of salary and personal emoluments for research and extension personnel, (c) national agricultural policies towards extension, with respect to the percent of the agricultural gross domestic product (AGDP) that is invested in agricultural extension.

Human Resources of Agricultural Extension

Adequate Number of Extension Personnel

Considering the present situation of agricultural extension personnel, Maalouf (1989, p. 185) mentioned that the numbers of extension workers in many countries are grossly inadequate in terms of current projected requirements such as one field worker for 1000 rain-fed or 500 irrigated farms. Supporting this view Swanson (1984, p. 6) mentioned that

"In most Third World countries, there was, and still is, a severe shortage of trained agricultural personnel" (p. 6).

Esman (1983) mentioned that "The principal task of contacting rural people and delivering these services falls on the field staff of government agencies. These field staff are often insufficient in number to maintain effective contact with the families they are expected to serve" (p.1). Brisset (1986) reported the shortage of extension personnel in the article: irrigation: the absolute imperative. van Apeldoorm and Turner (1984) reported that "the levels of extension contact and involvement with local development projects are low". Azrag (1987) reported that "in the Arab World there is a shortage of qualified weed scientists and extension workers."

Maalouf (1989) mentioned that "Inadequate numbers of well-qualified field, supervisory, and subject-matter staff are an almost universal problem" (p. 174). Brown (1983) reported that "the lack of subject-matter specialists (SMSs) is often the weakest link in the extension system and recommends the development of this group as a priority." Benor, Harrison, and Baxter (1984) also noted that "Usually, too few good SMSs are available to start with, so they must be used economically" (pp. 58-59).

Clark (1987) noted that "Human resource development is more than staff development—it's a programme that has as its goal the optimum match between the needs of the organization and the human resources available. This match isn't only based on current needs, but also on future needs of the organization" (p. 25).

Adequate Research and Extension Personnel by Gender

World Bank (1980, p. 3) argues, women make up 70 per cent of the agricultural labour force in many developing countries. They frequently weed, harvest, and store field crops; they sometimes comprise a majority of plantation labour. Yet new agricultural methods are usually taught to men. Cooperatives and other forms of self-help are usually dominated by men, and women find it hard to get credit on their own. Agricultural research has focused heavily on cash crops, controlled by men, and not on the subsistence crops grown by women. Supporting the same view Swaminathan (1985, p. 52) notes that:

> In many countries, and particularly in Africa, East and South-east Asia, women play major roles in farm and financial management.

> Moreover, many of their agricultural roles require considerable skills, for example, in seed selection, storage, and pre-germination testing, organic recycling, and the identification and control of pests, pathogens, and weeds. Nonetheless, many of the jobs undertaken in rural areas by women involve much drudgery and result in government policies, scientists, and manufacturers of agricultural machinery have generally tended to neglect women's technology needs. Women have also been generally bypassed by extension services. Their numbers in scientific and extension services are generally small. Consequently, they continue to remain as unskilled workers with uncertain income. *The challenge hence lies in integrating brain (technology), brawn (physical work), and bank (credit and other resources) in a mutually supportive manner so that both men and women can all play an active role in improving the productivity, profitability, stability, and sustainability of major farming systems* (p. 52).

Gabriel (1989) reported that "benefits from a male oriented extension service will be passed down to women have shown to be impracticable." Doorenbos, Haverkort and Jiggins (1988, p. 106) reported that "many extension services continue to face severe resource problems and it is common to find these used as an excuse for delaying action to reach women farmers: reaching women often is seen still as an effort requiring additional rather than a re-allocation of existing extension resources. Where revised work programming takes no account of the need to reach women farmers, changes introduced in the name of organizational efficiency may still fail to improve effectiveness. The fact is that most extension services are staffed wholly or predominantly by men who, for one reason or other, may have little contact with or incentive to reach women farmers" (p. 106). Information collected from 46 African countries (cited in Doorenbos, 1988) showed that only 3.4 per cent of trained government workers providing agricultural advice to rural people, were women.

Maalouf (1988) reported that "in the initial stages of extension programmes, women farmers prefer a female extensionist, on condition that she is competent in giving advice on agricultural subjects" (p. 10). Berger, Delacey and Mellencamp, (1984) reported that women extension workers should be incorporated into mainstream extension activities on an equal basis with men, not left to work with women. In addition both men and women extensionists should receive training to increase their awareness of the needs of women farmers. Finally, incentive systems should reward staff who are successful in reaching and assisting women farmers. McKee (1984) suggests that it is important "to involve women farmers and farm workers as well as female extension agents, in diffusing technologies for crops and tasks in which women predominate" (p. 602).

World Bank (1980) mentioned that "Nowhere are women adequately represented at higher levels of government administrations or parliaments. Professional training for women lags far behind that for men. Even at the village level, development workers–themselves mostly male-rarely seek to determine women's real needs and wishes" (p. 5).

Potash, (1985, p. 58) also mentioned that the extension agent's messages do get communicated by the headmen to the leaders of the neighbourhoods but it is not clear how much further the message goes. Certainly it does not reach women. Greater efforts must be made to reach women, and particularly older women, either through organized meetings, individual visits, or both. According to Belloncle (1989):

> Because of sociological factors it will usually be necessary to obtain the prior agreement of the men before the women can be included. And the women may first have to meet separately to express their own thoughts about their problems before a general meeting is held. Female specialists will be needed to enable true "bargaining" among all social groups (men, women, young people) within the village, so that the necessary modernization will be everyone's concern and will benefit all (p. 43).

Fortmann (1985) noted that:

> teaching staff at the training college for Agricultural Demonstrators (ADs) report that women are turned down who are better qualified than the men who are accepted. The official reason is that there is not enough dormitory space for women. Others argue that academically qualified women do not perform well in the field. There is no objective evidence to support this claim. Women are better able to work with the women farmers who head 25-30 per cent of the rural households in Botswana and, though subject to the short-term disability of pregnancy, they seem to have proved to be less likely to suffer the long-term disability of chronic alcoholism which troubles so many of their male colleagues. The issue of merit-based recruitment and any adjustments which it may entail needs to be considered for all cadres, but especially for ADs" (p.16).

The study of Gintings (1982) in Jawa reported that "in terms of sex, there was a significant difference in the empathic ability between male and female field extension workers." According to Boulding (1977), in some countries in Africa it is estimated that women do 70 per cent of the farm work. In such cases, Whyte (1986, p. 9) mentioned that women can be severely disadvantaged if they receive no assistance from the extension agents. Evaluating the Venda extension service, Bembridge (1989) reported that "12 of the 92 agricultural officer posts are filled by women. Because of the high preponderance of women carrying out farming operations, there is a need for greater numbers of women field workers" (p. 20).

Quality of Extension Personnel

Marshall (1930), mentioned that "Knowledge is our most powerful engine of production". Supporting the same view Roling (1988, p. 18), mentioned that "working through people as an extension communicator requires professional skill and insight which must be underpinned with knowledge". He also mentioned (1988), that "the introduction of innovation in a farming community is a strategic undertaking which has important socio-economic constraints" (p. 3). Spicer (1986) said *"Changing peoples' customs is an even more delicate responsibility than surgery"* (p.104). He further mentioned that:

> The administrator of a programme of technological change carries a heavier responsibility. Whenever he seeks to alter a people's way of life, he is dealing not with one individual, but with the well-being and happiness of generations of men and women. If his skill is poor and his judgement bad, he can destroy cooperative human relations and create hatreds that will affect unaccountable numbers of people. If, on the other hand, his skill is equal to the task, the possibility is open for creating cooperation where it did not exist and for bettering the lives of generations. The gravity of the responsibility need not overawe, but it must impress, each candidate in the profession with the necessity for the most careful training as a basis for every judgement.

On the basis of the Berlin survey, von Blanckenburg (1988, p. 27) mentioned that Bangladesh, Botswana, Kenya, Lesotho, Malaysia, Rwanda, the Sudan, the United Republic of Tanzania and Zambia, have unsatisfactory educational levels of extension staff. He further mentioned that staff education and training are more of problem in Africa than in Asia. Jones (1985) reported that "a shortage of adequately trained agricultural extension workers has been one of the limiting factors in establishing a viable agricultural extension service in the southern region of Sudan."

Reporting on the Venda agricultural extension services, Bembridge (1989, p. 20) mentioned, there are only three graduate level subject-matter specialists and administrators out of 112 staff members. Specialists posts are mostly filled by non-university graduates, who lack specialized knowledge and training. He further mentioned that there are considerable shortcomings in the technical competence of the extension staff.

Esman (1980, p. 1) noted that professionally qualified people, even when available, resist placement in rural areas; and when they are drafted into such assignments they often are so alienated from the majority of their rural publics and so ill-equipped to work in a resource-poor situation that they are unable to provide effective or responsive services.

Whyte (1986) stated that "We find many common deficiencies in knowledge and ability on the part of extension agents"(p. 5). In most developing countries, college education had been confined largely to persons coming from urban families. Men of such background who are employed as extension agents are largely a product of book learning. Relationships built upon such a foundation could hardly lead to constructive outcomes. Esman (1983, p. 1) mentioned that "All too often these technicians are indifferently trained, inadequately equipped, poorly paid, and under motivated". The Ministry of Agriculture in India (1988) expressed the same views in the findings of an Andhra Pradesh agricultural extension project. Peterson, Sands and Swanson (1989) reported that "less well qualified personnel significantly outnumber highly trained researchers in developing countries" (p. 61).

Lele (1975, p. 50) reported that there is lack of either a profitable technological package to meet the needs or no effective training of extension agents which makes much of extension's effort ineffective. World Bank (1985) noted that "An extension agent with just a secondary school education will not have the foundation necessary to understand farmers' problems and to teach them about increasingly complex forms of agricultural technology. As the quality of extension personnel improves, countries will need fewer staff because quantity can be substituted by quality" (p. 65).

According to Chaudhary and Al-Haj (1985, p. 169) a common barrier to effective extension work is the extension worker's lack of competence in such important areas as technical knowledge, farming skills, economic analysis, research procedures and communications ability. Describing the performance of extension workers in the Middle and Near East, Chaudhary mentioned that the relevance of the extension workers response depends upon his competence; *i.e.* the quality and quantity of knowledge adequate or sufficient for the purpose. In countries that are at the initial stages of their development, the extension worker's performance is more crucial. The less developed a country, the more competent should be the extension agent and the more the areas of his competency. In other words, in a developing country, with the majority of farmers being not able to read and write, farmers will depend entirely on the extension worker. The farmers in these countries lack access to various communication media and sources of information. If the extension worker can not provide the necessary and appropriate answers to a farmers questions, or if the agent gives wrong advice the farmer is going to suffer and the extension agent is going to lose his credibility.

Nagel and Schubert's (cited in Luning, 1984) mentioned that "the shortage of trained workers often inhibits all aspects of development performance. Large deficits in skilled labour power are particularly conspicuous in sub-Saharan Africa, where institution building has lagged much behind similar activities in South and South-east Asia. Yet even in the latter part of the world there exists a large disparity between theory and practice, often aggravated by 'Urban bias' in formal education".

According to Evenson (1988, p. 22), there is quality difference between the extension workers of low-income and advanced countries. Extension workers have received quite advanced training in most industrialized countries, but most extension workers in low-income countries had limited training. According to Benor (1987):

> Extension workers must be able to identify production constraints in the field and to either advise farmers how to overcome them or relay them accurately to research. To achieve this, extension workers at all levels need to be trained continuously to handle their particular responsibilities. *Only with professionally trained staff will credibility for extension be built within the farming community.* As well, extension workers must receive the basic physical and administrative resources and other support that are required to perform their functions" (p.138).

Adequate Training for Extension Personnel

Schultz (1971) in observing about the importance of human capital development in agriculture, paraphrased a line from William Faulkner's *Intruder in the Dust*: "The man without skill and knowledge leaning terrifically against nothing" (p. 47). Ames (1988) noted that "If the investment in trained manpower is limited or lacking, information may never be utilized" (p. 3). Roling (1988) mentioned that "Unfortunately, many extension professionals have been trained too narrowly and focus on technology and its introduction. Thus they miss many opportunities for helping their clients, as well as missing out on a lot of the exciting aspects of rural extension, and of intervention through communication" (p. 8).

Schultz (1981, p. 46) reported that, for developing highly productive economy only natural resources, physical capital and raw labour are not sufficient. Dynamics of development requires a wide array of human skill. In support of this view, he provided the example of the economic development of the early Mediterranean city-states, of the free cities of the northern Europe, and currently, of Hong Kong, Singapore, Denmark, Switzerland, and Japan.

Benor (1987) noted that "Just as extension is best done by a professional extension source, so are these functions best performed by specialized staff trained and able to work full-time in their specific fields" (p. 139). Oram, Zapata, Alibaruho, and Roy (1979) suggested that "Training must clearly receive very high priority in future planning since the shortage of sufficient trained staff is at least as great a constraint on research and extension in developing countries as is the shortage of funds" (p. 134). Clark (1987) mentioned that "The effectiveness of the educational programmes of extension depends on the abilities of its professional and volunteer staff. Continued professional development must become a priority, if, extension is to remain a viable source of information for a changing world" (p. 25).

Incentives for Extension Personnel

Incentive and motivation to extension personnel plays great role in the efficiency and effectiveness of extension services. von Blanckenburg (1984) mentioned that "The effectiveness of an extension service depends to a noticeable degree also on the morale and the motivation of its staff. In order to be efficient, the staff need acceptable working conditions, incentives, such as achievement-oriented pay and promotion opportunities and good staff management" (p. 27). Vengroff (1984) also reported the same views about the incentives system to extension agents in Upper Volta and Zaire.

Johnson's study (cited in Bramblet and Flewllen, 1976), notes that the "absence of professional incentives for those individuals who choose continuing education and public service as their vocation. Often the lack of these incentives are identified by comparison of the rewards systems existing for teaching and research faculties with those existing for extension faculty. Lower salaries, the lack of tenure and the absence of professional career streams are expressed as the discriminatory differences between the two systems" (p. 9).

The evaluation study of Bembridge (1989) confirmed that "The extension officer's own self-esteem, which is assumed to be a function of past success or failure, is an important motivational factor. Many of the middle level and junior extension officers interviewed saw themselves as having low status in the community. For example, they saw school teachers as having higher status than themselves. Possibly this was because they did not perceive their own role as adult educators. Reasons given for the lower status held were inferior salaries, lack of recognition, inadequate training and lack of transport" (p. 21).

Along with other incentives, better housing and other facilities to be provided to the extension personnel posted in the rural areas. von Blanckenburg (1984) mentioned that "Certainly the living conditions of field agents are often difficult. They stay in villages which may not have many of the facilities of modern civilization, such as electricity, running water, cultural events, or good schooling for their children. Many of them are separated from their families" (p. 27).

Due to the poor pay structure and less promotion opportunities new and qualified extension personnel prefer to join other organizations having better opportunities. von Blanckenburg (1984) mentioned that "In many countries extension field agents are on a rather low salary grade. Incentive payments are rarely provided in government services. In this respect the situation is often better in parastatals which can draw up their own staff and payment rules. Because of such hardships many services often lose their most capable officers who are lured away by the offer of better posts in private industry. This was, for instance, reported from the Ivory Coast and the Republic of Korea" (p. 27). According to World Bank (1985) "Salary scale and allowances were set by the civil services. The provision for improving salaries across the board or on the basis of merit was limited" (p. 61).

Amount of Work to Extension Personnel

The efficiency of extension personnel declines when they have to look after many educational and non-educational activities.

Hildebrand (1986, p. 5) noted that:

> "his effectiveness can be undermined by the scope of work he is expected to cover in tasks, territory, and population. In a study of the extension of high-yielding rice varieties in Tamil Nadu State of India, agents were responsible for supervising 30 or more other schemes, many of which had multiple operational components (one included promotion of five varieties, a loan programme, and fertilizer distribution). In addition, the extension agent was to submit a 21-page monthly report which could take up to a week to complete (Heginbotham 1975 pp. 107-108). Superiors freely demanded still other reports and soil samples from hundreds of randomly selected locations, setting 'targets' beyond any human capability to achieve (pp. 112-119)".

He further mentioned that " In neighbouring Andhra Pradesh State, a study found extension officers spending 19 to 44 per cent of their time in preparing and maintaining reports and returns, and that District Agricultural Officers had about 125 reports (weekly, monthly, quarterly, half-yearly, and annual) to submit" (Reddy 1981 p. 103).

Whyte (1986, p. 7) concluded that agricultural ministries should employ more agents so as to achieve more intensive relations between agents and farmers. Even if there were no other reasons to argue against this conclusion — and indeed there are — it is obvious that developing countries simply lack the money and the professional talent to provide for this more intense technical assistance relationship.

von Blanckenburg (1982) noted that "There is no doubt that the personal and material resources are often insufficient for the extension job to be done. The extension density *i.e.*, the number of farmers per extension officer-varies considerably"(p. 42). Referring the Berlin study he mentioned that extension density "lies between 300 and 3000 in the developing countries. There seems to be one accumulation between 500 and 1000 and another between 2500 and 3000". Citing this study Fortmann (1985, p. 17) noted that Assistant Community Development Officers appear to carry a heavier workload. According to the World Bank (1985):

> A common feature within national extension systems has been the excessive involvement of field-level extension personnel in the non-education aspect of technology transfer. For example in India, prior to the Bank's involvement, village extension workers had multi-purpose assignments within the Community Development Programme; they spent only a limited amount of their time on agricultural programmes and even less time on agricultural extension. In Kenya, Nigeria, and Turkey, agricultural extension workers helped run production campaigns that were heavily involved in the provision of inputs and other services. In Brazil, Morocco, and Thailand, administrative responsibilities diluted the role of extension personnel (p. 62).

It is clear from the literature that non-educational tasks are diverting considerable resources from extension's primary role. Re-allocating these resources to full-time educational responsibilities would increase extension's effectiveness in transforming traditional agriculture.

Financial Resources of Agricultural Extension

Adequate Financial Resources Allocated to Agricultural Extension

From its own experience and from the reviews of projects funded by other donor agencies, the World Bank has recognized that the potential benefits from agricultural projects can only be attained if an effective form of extension is in place. This recognition of the importance of extension resulted in an increase in projects involving extension from 1 per year in 1960s to 34 per year in the 1980s, and in Bank lending increasing from about US $1 million per year to more than US $200 million per year in 1989 (Hayward, 1990, p. 115). But the financial situation of agricultural extension in many countries is inadequate.

According to von Blanckenburg (1984) "Government financial resources for rural development in poor countries are notoriously scarce. In the majority of countries the effect of this has not been so severe on the actual number of extension staff but rather on the level of salaries and on finance available for equipment and its maintenance" (p. 28). He further mentioned that a number of reports/complaints that extension programmes seriously suffers from a shortage of funds. Esman (1980, p.1) noted that governments have been unable, for the most part, to afford to train and employ professional and technical personnel insufficient numbers or to establish sufficient facilities through which they could provide rural publics with conventional services.

Howell (1989) noted that:

> Extension investment in Asia was concerned with ensuring better utilization of existing staff and physical facilities-not with building up staff structures and facilities virtually from scratch, as in much of Africa. If Asian models are to be followed in Africa, priority will need to be given to increasing substantially the level of staffing both in the field and at the supervisory level. It will also mean an increase in research expenditure for work at research stations and on adaptive trials. The establishment of effective research, extension, and field trials involve a certain amount of capital outlay, but even more significant are commitments to recurrent cost. This necessarily means that in the long-term governments, rather than donors, will have to shoulder an increasing share of the costs" (p. 88).

According to Whyte (1986, p. 9), "There are many farmers who earn more money than do extension agents and many have as much education as the agents themselves. It is also mentioned (p. 7) that developing countries lack money and the professional talent to provide for more technical assistance relationships. Such a conclusion lead to recognize that the basic design of the relationship is faulty: if we think only in terms of a one-on-one relationship between the extension agent and the farmer, then it is impossible to develop a cost effective system of agricultural research and extension. We therefore, need to think of organizational strategies which will not only provide more useful information but will channel it more effectively and economically to those who need to use it.

Evenson (1988), reported that:

> Many low-income countries do not have the capacity to train agricultural scientists, and must incur high costs to train research workers and to purchase scientific equipments. These differences in cost, of course, reflect large differences in 'quality' of extension workers. In some low-income countries the available supply of extension workers with scientific training in agriculture is very limited. With programmes to expand extension systems rapidly, administrators have devised ways to build extension systems that can utilize the relatively untrained and unskilled field worker" (p. 22).

Howell (1989) noted that "It is now generally agreed that corrective action should consist of increased investment in extension (particularly in recurrent costs) linked to reforms in management and training" (p. 86).

Adequate Transportation Facilities to Agricultural Extension (Mobility)

For effective extension work, extension personnel require more mobility to contact clients. It is more important in those developing countries where the agent and farmers ratio is less and where agents have to cover more area and larger numbers of farmers in low density areas. This problem is not severe in industrially developed countries especially (European and North American countries) where nearly all extension workers have a car or easy access to transportation.

von Blanckenburg (1984, p. 29) mentioned that there are usually not enough vehicles and not enough funds for vehicle maintenance, including the cost of fuel. He further mentioned that one of the few organizational remedies available is the sharing of one vehicle by several officers going on duty tour. But there are very limited scope for this as the destinations are rarely close to each other.

von Blanckenburg (1984, p. 28) pointed out that in the interview-type survey, 27 per cent of the extension officers said that they had to operate within more than 100 villages and 41 per cent had to look after 10,000 or more farm families. On an average, only one official vehicle was available for seven extension workers. Therefore, the majority of staff remain office bound, and little impact can be expected of extension under such conditions. He mentioned that the situation is the same in Liberia, Mali, and Morocco.

Benor, Harrison and Baxter (1984) however, suggested that "Sub-division staff need to be highly mobile. Sub-divisional Extension Officers, Assistant SDEOs, Subject Matter Specialists, and Training Officers have to cover a large area regularly. They travel to the VEW training centres, farmers' fields, and research stations. For this reason, there should normally be a vehicle for each SDEO and Assistant SDEO, and for each SMS team" (p. 59). Fortmann (1985) noted that:

> Transportation difficulties compounded the large size of the extension areas. Many workers have only bicycles which are not practical where there is heavy sand or mud, or dangerous wild animals, or long distances. Although there is a government subsidy loan programme available, some cadres are not eligible and others do

not become eligible until long after the point of their likely demoralization. Such official transport as is available is frequently inefficiently utilized. The transportation issue must be addressed either by facilitating the acquisition of private transport or through the more efficient use of official transport" (p. 18).

Adequate Salary and Housing to Agricultural Extension Personnel

Fortmann (1985) reported that:

> Field officers generally feel that they are suffering from comparatively low salaries and a lack of training opportunities. When rewards do come, they are often in the form of promotions out of the field into administrative positions. Officers who wish to remain in the field have no real chance for advancement there. This has the effect of draining the best workers from the field, thereby perpetuating an overall low calibre of village extension work" (p. 19).

Bembridge (1989) noted that in Venda agricultural extension service, "both senior and junior staff were dissatisfied with residential accommodation, salaries and transport, and maintained there was a lack of direction and adequate training. These problems were generally substantiated in discussions with head office management" (p. 20).

Benor, Harrison and Baxter (1984) suggested that:

> "Arrangements for housing and office space should be made; additional houses may need to be constructed, since sub-divisional staff, like all extension staff, must reside in their area of jurisdiction. Sub-divisional staff, again like all others, should have an adequate budget for farm trials, training aids, field days, and other extension operations" (p. 59).

Referring the extension service in Botswana Fortmann (1985) reported that:

> "Extension staff feel strongly that housing is a right and not a privilege and that it should be provided before an extension worker is posted to an area. This is particularly a problem in the western areas where it is seldom possible to find private accommodation.

Even where housing is provided, it is not always satisfactory. Agricultural Demonstrator (AD) housing was frequently referred to as 'tin trunks' and the lack of toilets noted. One AD waited for months while the village squabbled over who should repair the roof which had blown off his house. An extension worker who is struggling to get the roof of his house repaired is unlikely to be giving full attention to his work. Finally, although they are both trained for the same length of time in the same institution, Assistant Community Development Officers (ACDOs) enjoy a comparatively good standard of living while ADs alternately swelter and freeze in their little 'tin trunks'. This leads to a sense of inequity. A standard housing policy for all cadres is necessary" (p.18).

National and International Agricultural Policies

National and international policies of agricultural research and extension play an important role in the development process. Dutia (1990) mentioned that "Achieving the long-run educational and development goals of extension requires continuity in political and financial support and the institutionalisation of the extension function. Extension can perform its role in development only when a favourable environment for its operation is created" (p. 41).

Nyiira (1987) noted that "policies play a major role in setting the stage for the implementation of scientific research and the application of science and technology". It is also mentioned by the author that "the lack of an integrated understanding of the impact of policies on research and its public use, often leads not only to poor policy structures but also inhibits national development" (p. 1).

Leonard (1977) mentioned that "extension services have the capacity to be considerably more effective instruments of agricultural development. A wide understanding of this potential can help to create political support for the reforms needed in the ministries of agriculture throughout the developing world" (p. 258).

Maalouf (1989) notes that:

> An indispensable starting point for effective and efficient extension work is a national extension policy incorporated in legislation. Such a policy specifies the objectives, clientele, funding arrangements, place of extension within the Government structure, its functions and linkages with other governmental and non-governmental services for agricultural and rural development" (p.173).

3

METHOD

This chapter is devoted to research design, population, instrument development, selection of indicators and measures, identification of the independent and dependent variables, and the collection and analysis of data.

Target Population

The target population was all member countries of the United Nations (UN) identified by the Food and Agricultural Organisation (FAO) of the United Nations and other sovereign non-member countries (which may be willing to respond to the survey) in six geographic regions of the world. The structured questionnaires were sent by the FAO to all countries through its resident representatives. Out of 159 member and 13 non-member countries, 117 (68%) countries responded to the survey. Two out of the 117 responding countries mentioned that they are not participating in this study because of administrative problems. The data of one country were sufficiently incomplete so these could not be included in the data base. However, the responses of 114 (66%) countries were considered to be acceptable. This group of respondents was considered to be sufficiently representative of the total population, so that findings that result from the analysis of data from 114 countries can be used to generalise to agricultural extension systems worldwide. The names of the responding countries are presented in Appendix B and response rate by region is given in (Figure 1) Table 1.

Instrumentation

The instrument used for this study was a four part questionnaire developed by a team of experts who have much experience in developing such

Table 1: Response Rate of the Responding Countries by Geographic Region

Geographic Region	*Population*	*Respondent*	*Response Rate*
Africa	48	31	64.58 %
Asia and the Pacific	33	26	87.78 %
Europe	35	14	40.00 %
Latin America and Caribbean	31	24	77.41 %
North America	3	3	100.00 %
The Near East	22	16	72.72 %
World	172	114	66.27 %

Note— The number of countries was counted from the *Europa World Year Book*, (1990, pp. 3-4).

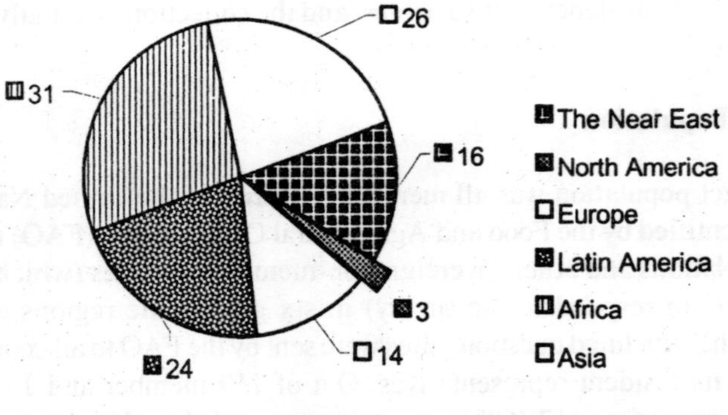

Fig. 1: Responding Countries Categorized by Geographic Region

instruments, including the researcher at the University of Illinois. The first part of the instrument was designed to collect secondary data to demographic and basic agricultural development indicators from the secondary sources. The second part of the instrument requested general information of agricultural extension organisations. The third part requested information on human resources, and the fourth part of the instrument requested information on financial resources available to agricultural extension organisations. This instrument was field tested and necessary corrections were incorporated in the final version. The last three parts of the instrument were translated into French and Spanish from the Original English version. The items (Questions) for all four parts of the instrument used in this study are compiled and presented in Appendix A.

Leading Indicators

Use of indicators is described and substantiated by Taylor (1980). He noted that:

> It has long been recognised that the integration of indicators into larger systems of measurement enhances analysis. These systems of indicators are developed with some regard to a theory about the phenomenon under study and tie together a series of indicators into a set of meaningful relationships with each other (p. 2).

Research Design

There are two major aspects of this study; the first is what is current status of human and financial resources of agricultural extension and the second is what is relationship between human and financial resources and primary agricultural development indicators. In order to assess the current status of human and financial resource of agricultural extension, descriptive statistics was used to analyse the human and financial resources data for 1988 and selected earlier years. Finally, multiple regression will be used to analyse the relationship between human and financial resources and agricultural development indicators.

Land (1981, p. 26) defined social indicators as "Constructs ... which are to assess the current state of, and to measure changes in, socio-economic conditions of life in contemporary societies". According to Morita-Lou (1985, p. 10) "the common thread underlying the various definitions is that indicators are tools used to assist in the assessment of some activity, action or consequences".

Deutsch (1980, p. 11) mentioned that "indicators ... tell us something about underlying conditions or processes and about their changes. Indicators things that are easy to observe and that are coupled or linked supposedly or in reality to other things that are harder to observe, but which we consider important". Deutsch further mentioned that "no single indicator can be trusted". Finally he stated (pp. 11-12) that "indicators are tools of discovery, not tools of proof. Only their cumulative weight can corroborate what we do or what we think".

Describing the utilization and impact of indicators, Morita-Lou (1985, p. 12) mentioned that "Utilization indicators are important for understanding the extent to which available resources and knowledge are effectively deployed towards stated objectives. Impact indicators are important for examining the actual influence produced in a particular context".

According to SPSS Inc. (1988, p. B-17), a leading indicator is a series that helps predict the values of another series one or more time periods later.

Considering the facts mentioned above, the agriculture development related predictor indicators were selected from the 'Socio-economic indicators relating to the agricultural sector and rural development', a publication of the Food and Agricultural Organisation of the United Nations (1984), 'Agricultural Development Indicators a Statistical Handbook' of the International Agricultural Development Services (IADS) (1980), World Development Report 1989 (World Bank), and from the 'International Directory of Agricultural Extension Organisations' (Swanson *et. al.* 1990).

Data Collection Procedures

The data on general information, human, and financial resources of agricultural extension organisations were collected from November 1988, through December, 1989 with the help of Food and Agricultural Organisation (FAO) of the United Nations. Copies of this instrument were sent in the appropriate language format to the FAO representative for each country. Instruments were then forwarded to all extension institution(s) in each country, with appropriate follow-up measures being employed. Completed instruments were returned to FAO, Headquarters in Rome and then forwarded to the Director, International Programme for Agricultural Knowledge System (INTERPAKS) at the University of Illinois, Urbana, USA. Since the researcher was involved in this FAO study from the outset, formal approval to use these data for this study was included in contract between the FAO and the university. Data about the human and financial resources of agricultural extension organisations from 1960 to 1988 were taken from this FAO data base. Data related to the agricultural development indicators of all responding countries from 1960 to 1988 have been collected by the researcher from secondary sources.

The basic agricultural development indicators as well as human and financial resource data from the last two parts of the instruments have been coded and recorded into two computerised data bases, using both individual countries and individual institutions as the unit of analysis. The Statistical Package of the Social Sciences (SPSS) was be used in carrying out the analysis.

Pilot Study

After designing the instrument with the help of the university experts and a similar study conducted by Swanson and Rassi (1980-81), the face validity was established by students and staff in the International Programme for Agricultural Knowledge Systems at the University of Illinois at Urbana-Champaign. Some modifications were made at this stage. Then the last three parts of the instrument were sent to the FAO head office, Rome for its pilot testing. After pilot testing, the necessary corrections were incorporated and revised instruments were sent to the FAO headquarters for data collection. The primary purpose of the pilot study was to determine that the appropriate items related to general information, and human and financial resources are included in the instrument.

Research Questions and Operational Hypotheses

To achieve the objectives of this study, specific research questions have been addressed using descriptive statistics. In addition, operational hypotheses were be tested; null hypotheses 1-3 were be tested at 0.05 level of significance. These research questions and hypotheses follow:

R.Q.1. What is the quantity (number) of existing human resources of agricultural extension organisations in six geographic regions, five economic groups, and four different levels of agricultural population of the world and their relationship with agricultural development?

Ho 1: There is no relationship between the quantity of human resources, as measured by the extension agent-farmer ratio, and selected agricultural development indicators.

Variables

Independent

Agent-farmers ratio*

(*Economically active population in agriculture divided by total extension personnel = Agent-farmer ratio.) These human resources data can also be broken down by gender (male, female) and type of position (administrative, subject-matter specialists, and field staff). Further analysis of these sub-categories of data will be considered.

Dependent

Agricultural development indicators
Average annual growth rate of agriculture (per cent)
Agricultural production index
Daily calorie supply
Daily grams of protein
Fertilizer consumption
Density of tractors

Control

Percentage of literacy
Number of radios per 1000 households
Number of televisions per 1000 households
Number of newspapers per 1000 households

Analysis

Multiple regression between agent-farmers ratio and agricultural development indicators.

R.Q.2. What is the quality of existing human resources of agricultural extension organisations in six geographic regions, five economic groups, and four different levels of agricultural population of the world and its relationship with agricultural development?

Ho 2: There is no relationship between the quality of agricultural extension personnel, as measured by educational qualifications, and selected agricultural development indicators.

Variables

Independent

Educational qualification*

High school	=	12 years
Intermediate	=	14 years
First university degree	=	16 years
Post-graduate	=	18 years

(*Quality of extension personnel is equated with educational qualification in this study and measured as composite variable representing the average educational level of the total extension staff, where secondary education = 12 years, intermediate education = 14 years, first university degree = 16 years, and post-graduate degree = 18 years. The percentage of staff at each level were multiplied by these respective years of education and then these numbers were summed to create an average educational level for the extension staff in each country, ranging somewhere between 12 and 18 years.)

Dependent

Agricultural development indicators
Average annual growth rate of agriculture (per cent)
Agricultural production index
Daily calorie supply
Daily grams of protein
Fertilizer consumption
Density of tractors

Control

Percentage of literacy
Number of radios per 1000 households
Number of televisions per 1000 households
Number of newspapers per 1000 households

Analysis

Multiple regression between quality of extension personnel and agricultural development indicators.

R.Q.3. How much finance is allocated to agricultural extension organisations in six geographic regions, five economic groups, and four different levels of agricultural population of the world and its relationship with agricultural development?

Ho 3: There is no relationship between the financial allocation to agricultural extension organisations, as measured by the average extension expenditure for each economically active farmer, and selected agricultural development indicators.

Variables

Independent

Extension expenditure per economically active farmer*

(*Note*— These data are also broken down by salary, programme and capital maintenance costs.)

*Extension expenditure in each country was be divided by the economically active population in agriculture to obtain the average expenditure (in US Dollars) per farmer.

Dependent

Agricultural development indicators
Average annual growth rate of agriculture (per cent)
Agricultural production index
Daily calorie supply
Daily grams of protein
Fertilizer consumption
Density of tractors

Control

Percentage of literacy
Number of radios per 1000 households
Number of televisions per 1000 households
Number of newspapers per 1000 households

Analysis

Multiplicative effect and Multiple regression between expenditure per farmer and agricultural development indicators.

R.Q.4. Given past trends and current levels, plus the projected number of economically active population in agriculture and specific assumptions such as:

* extension personnel trend,
* existing agent-farmers ratio,
* existing ratio of agent and area covered,
* current literacy rate and trends,

* existing use of mass media, and
* constrained level of system growth (*i.e.* 10 per cent per year)

What level of human resources will be required (or are possible) by the agricultural extension organisations, in six geographic regions, five economic groups, and four different levels of agricultural population of the world.

Analysis

Prediction and multiplicative effect analysis

R.Q. 5. Given the current level of extension expenditure and different assumptions such as:

* expenditure trends,
* existing expenditure per farmer,
* expenditure per agent, and
* constrained level of economic expansion (*i.e.* 10 per cent per year).

How much finance will be required (or what expenditure level seems possible) for agricultural extension organisations, in six geographic regions, five economic groups, and four different levels of agricultural population of the world?

4

RESULTS OF THE STUDY

This chapter is devoted to presenting the result of this study based on an analysis of the data. These results are presented in four parts according to the objectives of the study. The data analyzed under objectives 1-3 were presented by geographic region and economic group. In addition, the relationship of each independent variable was tested with the corresponding dependent variables, using the Chi-Square test for four null hypotheses. These data were further analyzed using one way analysis of variance (ANOVA) of the log of the computed values for each variables to see the multiplicative effect on them.

The objectives of the study were: (a) estimate and compare the quantity of existing human resources of national agricultural research and extension systems, categorised by geographic region and economic groups, and to determine their relationship with selected agricultural development indicators, (b) to estimate and compare the quality of existing human resources of agricultural extension systems, categorised by geographic region and economic groups and to assess the relationship of these measures with selected agricultural development indicators, and (c) to assess the relationships of these selected agricultural development indicators with specific numbers of human and financial resources in national agricultural research and extension systems. Four agricultural development indicators were identified and used in this analysis, including: total yield of cereals in kilograms (kg)/hectare (ha), farmer income, agricultural income/ha of arable land and total fertilizer consumption in kilograms of N, P, K for each hectare of arable and permanent cropland. Data of these four dependent variables are presented in Appendix C categorised by geographic region and economic group. Also, under this fourth objective, four null hypotheses were tested using multiple regression analysis at .05 level of significance.

The analytical techniques used included descriptive statistics, bivariate analysis (Chi-Square test and correlation), and multivariate (multiple regression) analysis. It should be noted that these analyses were carried out using two data sets: all responding countries and those countries with per capita GDP of less than U.S. $ 5,000 (*i.e.* the "developing" countries). The purpose of this dual analysis was to determine if the same findings would hold for the developing countries alone, as contrasted with the worldwide data set including both the industrially developed and developing countries.

Demographic Information

The initial plan was to analyze the data using three major groups: geographic regions, economic groups based on per capita GDP and the relative proportion of agricultural population in each country. However, after preliminary analysis of data, it was realized that the information in many tables were either redundant or very similar, especially between the proportion of agricultural population and the economic group based on per capita income. Therefore, the data are presented only by geographic region and economic group throughout this chapter.

Table 2 presents the responding countries by geographic region and economic group. Among the economic group, the largest number of countries were in lower middle income (36%), followed by low income group (32.5%), high income group (21.9%) and upper middle income group (9.6%). Among the regions, the largest number of countries were from Africa (27.2%) followed by Asia (22.8%), Latin America (21.1%), the Near East (14%), and North America (2.6%).

Table 3 presents the responding countries by geographic region and by the percentage of the population in agriculture. The largest agricultural population countries were found in Africa (27.2%) followed by Asia (22.8%), Latin America (21.1%), and the Near East (14%). The agricultural population in North America and Europe was found to be very small in comparison to the other regions of the world.

Given this overview of the general characteristics of the responding countries, we will now proceed with an analysis of these data, following the four major objectives set out for this study. First, data on the number or quantity of extension personnel will be analyzed, given in objective one, which follows:

Objective 1: *To estimate and compare the quantity of existing human resources in national agricultural research and extension systems,*

categorised by geographic region and economic group and to determine their relationship with selected agricultural development indicators.

Table 2: Total Responding Countries, Categorised by Geographic Region and Economic Group

Geographic Region	*Economic Group (PCI*)*				*Region Total%*
	Low Income	*Lower Income*	*Upper Middle Income*	*Higher Income*	
Africa	21	9	1	-	31 (27.2)
Asia	11	11	1	3	26 (22.8)
Europe	-	-	2	12	14 (12.3)
Latin America	2	15	7	-	24 (21.1)
North America	-	-	-	3	3 (2.6)
The Near East	3	6	-	7	16 (14.0)
Group Total	37	41	11	25	114
(%)	(32.5)	(36)	(9.6)	(21.9)	(100.0)

* PCI = Per Capita Income (Economic groups are categorised by the per capita Gross Domestic Product (GDP) of each country in 1988, based on four groups: Low Income = < U. S. $ 500, Lower Middle Income = U. S.$ 500-2,000, Upper Middle Income = U.S. $ 2,000-5,000, and High Income = > U. S. $ 5,000. N = 114.

Table 3: Total Responding Countries, Categorised by Geographic Region and the Percentage of the Population in Agriculture

Geographic Region	*Agriculture Population Group**				*Region Total %*
	Low-Ag Population Group	*Lower-Mid Ag. Pop. Group*	*Upper-Mid Ag. Pop. Group*	*High Ag. Pop. Group*	
Africa	-	4	3	24	31 (27.2)
Asia	4	5	7	10	26 (22.8)
Europe	14	-	-	-	14 (12.3)
Latin America	5	13	5	1	24 (21.1)
North America	3	-	-	-	3 (2.6)
The Near East	6	4	5	1	16 (14.0)
Group Total	37	41	11	25	114
(%)	(32.5)	(36)	(9.6)	(21.9)	(100.0)

* Agricultural Population groups are categorised by the percentage of population in agriculture in four groups, where low agriculture population = < 20 per cent, Lower middle agricultural population = 20-30 per cent, Upper agricultural population groups = 40-59 per cent and high agricultural population group > 60 per cent.

The effectiveness of agricultural research and extension systems depends largely upon the number and quality of their personnel, and how well they are supported technically and financially, as well as opportunities for career advancement and other professional rewards. Only the quantity of researchers and extension workers is discussed in this objective. Without adequate numbers of research and extension personnel, it is expected that agricultural research and extension will be constrained in planning and implementing effective research and extension programmes, including technology transfer activities. The quantity of research and extension personnel required, depends on many factors, including the size of the economically active population in agriculture, the size of the country in terms of arable land, the overall literacy rate of the rural population in each country, the density of radio and television sets, etc. Tables 4 and 5 present basic information from the responding countries regarding these basic demographic data, including the total number of researchers and extension personnel, grouped by six regions.

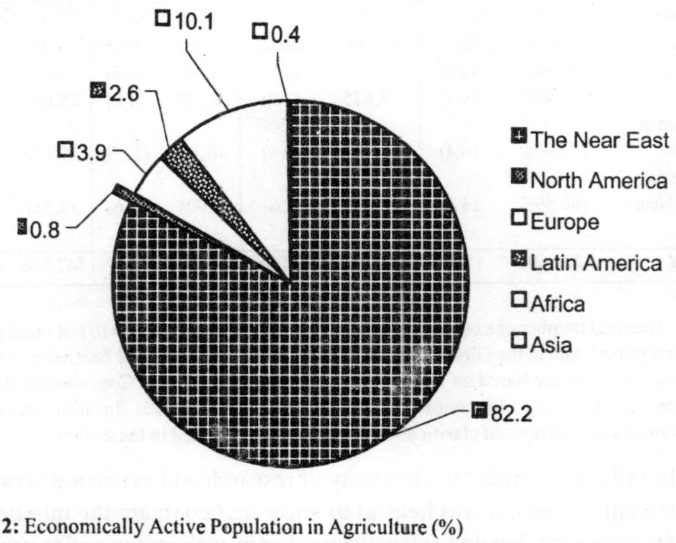

Fig. 2: Economically Active Population in Agriculture (%)

The purpose of Table 4 is to present an overview of the number of researchers and extension personnel in the responding countries for the various geographic regions, in relation to the general and agricultural production. It is evident from Table 4 that highest population numbers for the 114 countries included in this study were found in Asia (66.4%), followed by Latin America (9.3%), Africa (8.6%), North America (6.4%), Europe (4.9%), and the Near East (4.4%). The economically active population in agriculture was also highest in Asia (82.2%) (Figure 2), followed by Africa (10.1%), Latin America (3.9%), the Near East (2.6%), Europe (0.8%), and North America (0.4%). More than 22.5 per cent of the total researchers were found in European and North American countries and 77.5 per cent researchers were from the rest of the regions. The total number of extension personnel was highest in Asia with 390,191 (71.9%), followed by Africa (11%). The remaining extension personnel (about 17%) were found in the rest of the regions.

Table 4: Total Population, Economically Active Population in Agriculture, Total Research and Total Extension Personnel by Geographic Region

Region	*Total Population*		*Economically Active Population in Agriculture*		*Total Research Personnel*		*Total Extension Personnel*	
	(000)	%	(000)	%		%		%
Africa	364,652	(8.6)	100,501	(10.1)	3,334	(2.8)	59,676	(11.0)
Asia	2,804,946	(66.4)	814,487	(82.2)	71,902	(60.3)	390,191	(71.9)
Europe	206,885	(4.9)	7,457	(0.8)	10,487	(8.8)	15,127	(2.8)
Latin America	391,455	(9.3)	38,625	(3.9)	9,325	(7.8)	28,851	(5.3)
North America	272,068	(6.4)	3,531	(0.4)	16,314	(13.7)	15,170	(2.8)
The Near East	186,596	(4.4)	26,015	(2.6)	7,904	(6.6)	33,551	(6.2)
Total	4,226,602	(100)	990,616	(100)	119,226	(100)	542,566	(100)

Note– The total number of extension personnel presented in this study will not match with earlier reported data in the Global Consultation Report on Agricultural Extension (1990), because these data are based on 114 countries and the earlier Global Consultation Report include data for only 113 countries (see Swanson, *et al.* 1990). In addition, some supplementary changes and clarification were also incorporated in these data.

In order to compare the intensity of research and extension services in different regions, it was helpful to know and compare the number of farmers per researcher and extension worker in each region and to present data on other potential sources of agricultural information. Table 5 presents information on the number of farmers per researcher and per extension worker along with other demographic data, including the percentage of literacy, and the number of radio and television sets per 1000 households.

The level of literacy, the number of radio and television per 1000 household, and number of farmers per extension worker are all important means of increasing farmer access to agricultural information, skills and knowledge. If the literacy percentage and density and intensity of radio, television sets, and extension workers is relatively high, than farmers have greater access to agricultural technology, as well as to extension and education services. However, it is evident from Table 5 that the literacy percentage, the number of radios and television sets per 1000 household is many times less in Africa, Asia, Latin America and the Near East in comparison to Europe and North America. The number of farmers per research and extension worker is also considerably higher (*i.e.* a lower intensity of extension services per farmer) in Asia, Africa, Latin America and the near East, in contrast with Europe and North America. If Agricultural development is associated with increased intensity of research and extension services, then the number of researchers and extension workers in Asia, Africa, Latin America, and the Near East would need to be strengthened. This assumption or premise is tested using Chi-Square test in objectives 1-3 and through multiple regression analysis in objective four.

Table 5: Literacy, Radio, Television, and Farmers per Researcher and Extension Worker by Geographic Region

Region	*Mean Level of Literacy (%)*	*Mean No. of Radio/1000 Household*	*Mean No. of TV/1000 Household*	*No. of Farmers/ Extension Worker*	*No. of Farmers/ Researcher*
Africa	35	119	25	1684	30144
Asia	66	321	98	2087	11328
Europe	97	656	348	493	711
Latin America	78	360	109	1338	4142
North America	99	1441	740	233	216
The Near East	55	317	193	775	3219
Region Mean	63	346	150	1826	8308

Note— The literacy, Radio, and T.V. figures presented in this table are the mean numbers for each region, rounded to one unit if the number was more than 0.5. Literacy data were taken from The World Factbook 1989; Radio and T.V. data were taken from the UNESCO *Statistical Yearbook 1987*.

Extension Personnel

Table 6 presents the number of extension personnel by region and economic group. It indicates that the number of extension personnel in low income, lower middle income and upper middle income countries and in Africa,

Asia, Latin America, and the Near East were much higher than the number identified in Europe and North America. Among the regions, Asia has more than 71 per cent of all extension personnel worldwide (Figure 3) followed by Africa with 11 per cent. The rest of the regions have between 2 per cent and 7 per cent of all extension personnel identified in this study. Among the economic groups, low income countries have highest percentage of extension personnel (68.5%), followed by lower middle income group with about 21.45 per cent. The upper middle and high income group have a total of about 10 per cent of all extension personnel worldwide.

Table 6: Total Number of Extension Personnel by Geographic Region and Economic Group

Geographic Region	*Economic Group (PCI)*					
	Low Income	*Lower Middle Income*	*Upper Middle Income*	*Higher Income*	*Region Total*	*(%)*
Africa	43,641	14,735	1,300	-	59,676	11.0
Asia	325,914	44,420	7,979	11,878	390,191	71.9
Europe	-	-	2,162	12,965	15,127	2.8
Latin America	1,016	26,194	1,641	-	28,851	5.3
North America	-	-	-	15,170	15,170	2.8
The Near East	1,129	31,021	-	1,401	33,551	6.2
Group Total	371,700	116,370	13,082	41,414	542,566	
(%)	68.51	21.45	2.41	7.63		100.0

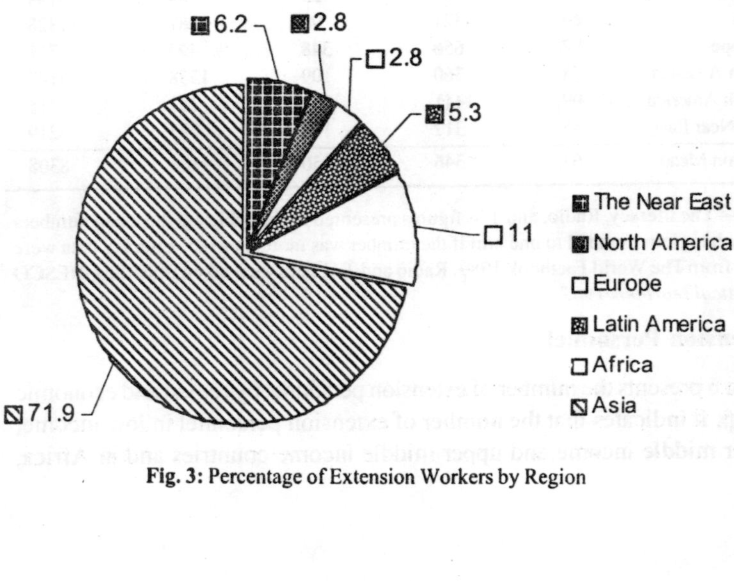

Fig. 3: Percentage of Extension Workers by Region

Number of Farmers per Extension Worker

Table 7 presents the number of farmers per extension worker (F/EW) by geographic region and economic group. It is evident from Table 7 that low income countries have a higher number of farmers per extension worker, or a lower intensity of extension services, than the high income countries. The number of farmers per extension worker in the low income group was found to be about 2,326 in comparison to high income countries where there were only 395 farmers per extension worker. The number of farmers per extension worker in the lower middle income group was 746, in comparison to the high income group with 396 farmers per extension worker. In other words, the number of farmers per extension worker among the economic groups ranged from 2,326 in low income groups to 396 in high income groups. This finding indicates the general trend, that the poorer country, the great number of farmers per extension worker. The number of farmers per extension worker was lowest in North America (233), followed by Europe (493) (Figure 4). The number of farmers per extension worker in rest of the regions were very high in comparison to North America and Europe.

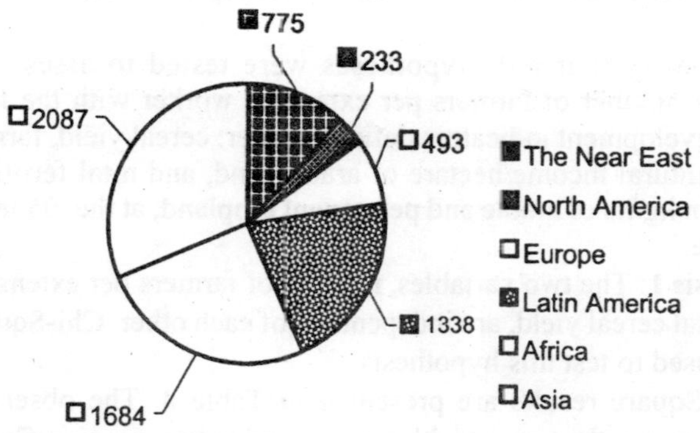

Fig. 4: Number of Farmers per Extension Worker by Region

The purpose of Table 4 is to present an overview of the number of researchers and extension personnel in the responding countries for the various geographic regions, in relation to the general and agricultural production. It is evident from Table 4 that highest population numbers for the 114 countries included in this study were found in Asia (66.4%), followed by Latin America (9.3%), Africa (8.6%), North America (6.4%), Europe (4.9%), and the Near East (4.4%). The economically active population in agriculture was also highest in Asia (82.2%) (Figure 2), followed by Africa.

Table 7: Number of Farmers per Extension Worker by Geographic Region and Economic Group

Geographic Region	*Economic Group (PCI*)*				*Region*
	Low Income	*Lower Middle Income*	*Upper Middle Income*	*High Income*	
Africa	1,947	960	1,061	—	1,684
Asia	2,371	715	609	419	2,087
Europe	—	—	681	461	493
Latin America	1,883	1,346	9,269	—	1,339
North America	—	—	—	233	233
The Near East	4,195	625	—	1,356	775
Group	2,326	746	1,751	396	1,826

Note— The number of farmers per extension worker presented in this table was computed directly by dividing the total number of farmers given in Appendix D (Table D-3) by total number of extension workers given in Table 6, regardless of the number of countries. Therefore, the row and column totals of this table, do not add up to the total.

The following four null hypotheses were tested to assess the relationship of number of farmers per extension worker with the four agricultural development indicators defined earlier: cereal yield, farmer income, agricultural income/hectare of arable land, and total fertilizer consumption in kg/ha of arable and permanent cropland, at the .05 level of significance.

Hypothesis 1: The two variables, number of farmers per extension worker and total cereal yield, are independent of each other. Chi-Square analysis was used to test this hypothesis.

The Chi-Square results are presented in Table 8. The observed relationship between the two variables was found to be very significant, $\chi^2 (4, N = 99) = 17.32, p<.01$, therefore, the null hypothesis was rejected. In practical terms, these findings indicate that as there are fewer number of farmers per extension worker or as extension intensity per farmer increases, then total cereal yield (kg/ha) also increases. When this same analysis was repeated only for the developing countries (per capita GDP of less than U.S. $ 5,000) included in the data set, the observed relationship between the two variables also was found to be very significant χ^2 (4, N = 62)=13.65, $p< .01$. The minimum expected frequency of one cell out of 9 cells was <5 (*i.e.* 4.88), but it is acceptable because it is < 25 per cent. Therefore, the null hypothesis was rejected, further confirming the expected relationship.

The results of multiplicative effect are presented in Table 9 indicates that fewer number of farmers (<600) per extension worker increases total cereal yield more than 42 per cent over the overall average ($e^{7.5018}$) 1811.29. Having 600-1,400 farmers per extension worker increases total cereal yield 5.51per cent, and having >1,400 farmers per extension worker decreases total cereal yield 29.37per cent.

Table 8: Frequency of Countries by the Number of Farmers per Extension Worker and Total Cereal Yield

No. of Farmers/ Extension Worker	*Total Cereal Yield (kg/ha)*			*Row Total*	*%*
	<1,400	1,400-2,500	>2,500		
< 600	5	9	15	29	29.2
600-1,400	9	16	10	35	35.4
> 1,400	19	12	4	35	35.4
Column Total	33	37	29	99	
(%)	33.3	37.4	29.3		100.0

Note— *N* Missing = 15, $\chi^2 = 17.325$, $df = 4$, $p=.00167$, MEF = 8.49.

Table 9: Multiplicative Effect of Number of Farmers per Extension Worker on Cereal Yield

No. of Farmers/ Extension Worker	*Means*	*SD*	*Effect on Transformed Variable*	*Percentage Increased*	*P*
< 600	7.8567	.6296	.3549	42.60	
600-1,400	7.5554	.5458	.0536	5.51	}.0000
> 1,400	7.1541	.5828	–.3477	–29.37	
	7.5018	.6449			

Hypothesis 2: The two variables; number of farmers per extension worker and farmer income, are independent of each other. Chi-Square analysis was used to test this hypothesis.

The Chi-Square results are presented in Table 10. The observed relationship between the two variables was found to be very significant, χ^2 (4, $N = 85$) = 29.20, $p < .0001$, therefore, the null hypothesis was rejected. In other words, this finding indicates that as there are fewer number of farmers per extension worker or as extension intensity per farmer increases, then farmer income also increases. When this same analysis was repeated only for the developing countries where per capita GDP of less than $ 5,000, then this relationship continued to hold with χ^2 (4, N = 62) = 14.25, $p < .01$.

The results of multiplicative effect are presented in Table 11 indicates that fewer number of farmers (<600) per extension worker increases farmer income more than 676 per cent over the overall average ($e^{6.3484}$) 751.57. Having 600-1,400 farmers per extension worker decreases farmer income 58.57 per cent, and having >1,400 farmers per extension worker decreases farmer income 67.7 per cent.

Table 10: Frequency of Countries by Number of Farmers per Extension Worker and Farmer Income

No. of Farmers/ Extension Worker	*Farmer Income (US$)*			*Row Total*	*%*
	< 225	225-1,500	> 1,500		
< 600	1	8	18	27	31.8
600-1,400	15	12	4	31	36.4
> 1,400	15	8	4	27	31.8
Column Total	31	28	26	85	
(%)	36.5	32.9	30.6		100.00

Note— N Missing = 29, χ^2 = 29.20, $df = 4$, $p = .0001$, MEF = 8.259.

Table 11: Multiplicative Effect of Number of Farmers per Extension Worker on Farmer Income

No. of Farmers/ Extension Worker	*Mean*	*SD*	*Effect on Transformed Variable*	*Percentage Increased*	*p*
< 600	8.3980	1.5148	2.054	676.48	
600-1,400	5.4672	1.8593	–.881	–58.57	}.0000
> 1,400	5.2784	2.8282	–1.07	–67.70	
	6.3484	2.5248			

Hypothesis 3: The two variables; number of farmers per extension worker and agricultural income, are independent of each other. Chi-Square analysis was used to test this hypothesis.

The Chi-Square results are presented in Table 12. The observed relationship between the two variables was found to be very significant, χ^2 (4, N = 85) = 15.085, p<.01, therefore, the null hypothesis was rejected. In other words, these findings indicate that as there are few number of farmers per extension work, the agricultural income per hectare arable land also increases. However, when this same analysis was repeated only for the developing countries where (per capita GDP of less than US $ 5,000), this relationship did not hold. Since the χ^2 (4, N = 62) = 3.222, p<.05, MEF = 5.210, then the null hypothesis could not be rejected at the .05 level of significance for the developing country data set.

Table 12: Frequency of Countries by the Number of Farmers per Extension Worker and Agricultural Income

No. of Farmers / Extension Worker	*Agricultural Income /ha (US$)*			*Row Total*	*%*
	< 525	525-1,500	> 1,500		
< 600	8	8	11	27	31.8
600-1,400	19	9	3	31	36.4
> 1,400	19	6	2	27	31.8
Column Total	46	23	16	85	
(%)	54.1	27.1	18.8		100.00

Note— N Missing = 29, χ^2 = 15.085, df = 4, p = .00453, MEF = 5.082.

The results of multiplicative effect are presented in Table 13 indicates that fewer number of farmers (<600) per extension worker increases agricultural income more than 256 per cent over the overall average ($e^{5.8230}$) 337.98. Having 600-1,400 farmers per extension worker decreases agricultural income more than 33 per cent, and having >1,400 farmers per extension worker decreases agricultural income 56.74 per cent.

Table 13: Multiplicative Effect of Number of Farmers per Extension Worker on Agricultural Income

No. of Farmers/ Extension Worker	*Mean*	*SD*	*Effect on Transformed Variable*	*Percentage Increased*	*P*
< 600	7.0937	1.3500	1.2707	256.33	
600-1,400	5.4211	2.4518	–.4019	–33.10	}.0036
> 1,400	4.9850	2.7235	–.8380	–56.74	
	5.8230	2.4028			

Hypothesis 4: The variables; number of farmers per extension worker and total fertilizer consumption in kg/ha arable and permanent cropland are independent of each other.

The Chi-Square results are presented in Table 14. The observed relationship between the two variables was found to be very significant, $\chi^2 (4, N = 98) = 22.978, p<.001$, therefore, the null hypothesis was rejected. In practical terms, these findings indicate that as there are few number of farmers per extension work or as extension intensity per farmer increases, then the total fertilizer consumption (kg/ha) also increases. When this same analysis was completed for only the developing countries (per capita GDP of less than US $ 5000), this relationship continued to hold with $\chi^2 (4, N = 77) = 19.268, p<.001$, hence the null hypothesis was rejected.

Table 14 : Frequency of Countries by the Number of Farmers per Extension Worker and Total Fertilizer Consumption

No. of Farmers / Extension Worker	*Total Fertilizer Consumption (kg/ha)*			*Row Total*	*%*
	< 25	25-100	> 100		
< 600	5	10	17	32	32.6
600-1,400	9	14	10	33	33.7
> 1,400	21	9	3	33	33.7
Column Total	35	33	30	98	
(%)	33.7	37.7	30.6		100.00

Note— *N* Missing = 16, χ^2 = 22.978, $df = 4$, $p = .00013$, MEF = 9.796.

The results of multiplicative effect are presented in Table 15 indicates that fewer number of farmers (<600) per extension worker increases total fertilizer consumption (kg/ha) arable and permanent cropland more than 153 per cent over the overall average ($e^{2.0237}$) 7.57. Having 600-1,400 farmers per extension worker increases total fertilizer consumption 12.36 per cent, and having >1,400 farmers per extension worker decreases fertilizer consumption 64.84 per cent.

Arable Land per Extension Worker

Another perspective on the importance of extension personnel concerns the average area of arable land that is to be covered by each extension worker. This ratio varies from country to country depending on the total arable land area and the number of extension personnel in each country.

Table 16 presents the number of hectares of arable land per extension worker, by geographic region and economic group.

Table 15: Multiplicative Effect of Number of Farmers per Extension Worker on Fertilizer Consumption

No. of Farmers/ Extension Worker	*Mean*	*SD*	*Effect on Transformed Variable*	*Percentage Increased*	*P*
< 600	2.9527	1.3874	.9290	153.20	
600-1,400	2.1402	1.4927	.1165	12.36	}.0000
> 1,400	.9783	1.9830	−1.0454	−64.84	
	2.0237	1.8175			

Table 16: Arable Land (ha) per Extension Worker by Geographic Region and Economic Group

Geographic Region	*Economic Group (PCI*)*				*Region*
	Low Income	*Lower Middle Income*	*Upper Middle Income*	*High Income*	
Africa	1,631	1,909	5,362	-	1,781
Asia	994	567	250	4,333	1,032
Europe	-	-	3,285	2,364	2,496
Latin America	1,019	1,747	57,501	-	4,892
North America	-	-	-	15,411	15,411
The Near East	8,010	1,000	-	1,148	1,242
Group	1,090	1,118	8,441	7,667	1,776

Note— Arable land per extension personnel presented in this table is computed directly by dividing the arable land given in Appendix D (Table D-4) by total number of extension workers given in Table 6, regardless of the number of countries. Therefore, the row and column total of this table do not add up to the total.

It is evident from Table 16 that among the different regions, more arable land was covered per extension worker in North America (15,411 ha) as in contrast to Asia (1,032 ha) (Figure 5). This finding respects the fact that farm sizes are much larger in North America, in comparison to other regions. Among the economic groups, arable land per extension worker was highest (8,441 ha) in upper middle income group and lowest (1,090 ha) in the low income group. The following four null hypotheses were tested to assess the possible relationship between the area of arable land per extension worker with the four agricultural development indicators; cereal yield, farmer income, agricultural income and total fertilizer consumption at the .05 level of significance.

Hypothesis 1: The two variables; hectares of arable land per extension worker and total cereal yield, are independent of each other. Chi-Square analysis was used to test this hypothesis.

The Chi-Square results are presented in Table 17. The observed relationship between the arable land per extension worker and cereal yield was not found to be significant, χ^2 (4, N = 82) = 6.729, p<.15, therefore, the null hypothesis was accepted. In practical terms, this finding indicates that there does not appear to be a relationship between arable land (ha) per extension worker and cereal yield. When this analysis was repeated only for the developing countries data set (per capita GDP of less than U.S. $ 5,000), the observed relationship between the two variables was found to be not significant, χ^2(4, N= 82) = 7.737, p> .10, but it was approaching significance at the .1 level. Consequently, the null hypothesis was accepted.

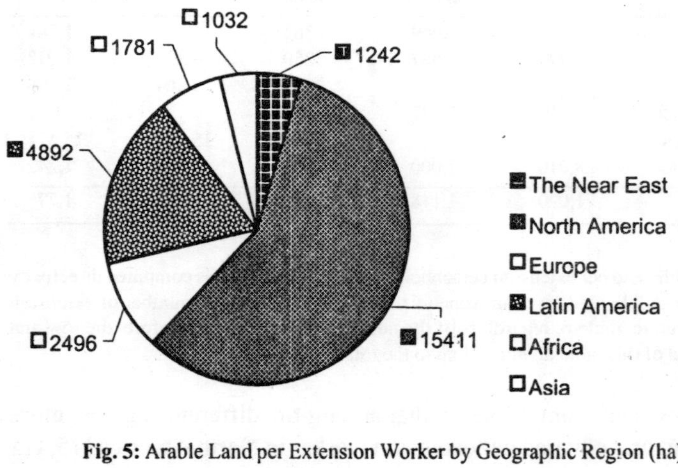

Fig. 5: Arable Land per Extension Worker by Geographic Region (ha)

The results of multiplicative effect are presented in Table 18 indicated that fewer hectares of land (<550) per extension worker was related to increase total cereal yield kg/ha of 15.94 per cent over the overall average ($e^{7.5114}$) 1828.77. Having 550-2,000 hectares of arable land per extension worker was related to decrease total cereal yield of 1.79 per cent, and having >2,000 hectares of arable land per extension worker was related to decrease cereal yield (kg/ha) about 9.62 per cent.

Table 17: Frequency of Countries by Number of Hectares of Arable Land per Extension Worker and Total Cereal Yield

Arable Land (ha)/ Extension Worker	*Total Cereal Yield (kg/ha)* < 1,400	1,400-2,500	> 2,500	*Row Total*	*(%)*
< 550	7	9	14	30	29.7
550-2,000	12	12	9	33	32.7
> 2,000	14	17	7	38	36.6
Column Total	33	38	30	101	
(%)	32.7	37.6	29.7		100.0

Note— *N* Missing = 13, χ^2 = 6.729, $df = 4$, $p = .15089$, MEF = 8.911.

Table 18: Multiplicative Effect of Arable Land per Extension Worker on Cereal Yield

Arable Land (ha)/ Extension Worker	*Mean*	*SD*	*Effect on Transformed Variables*	*Percentage Increase*	*P*
< 550	7.6593	.6383	.1479	15.94	
550-2,000	7.4933	.6660	–.0181	–1.79	}.0032
> 2,000	7.4933	.6404	–.0181	–9.62	
	7.5114	.6499			

Hypothesis 2: The two variables; hectares of arable land per extension worker and farmer income, are independent of each other. Chi-Square analysis was used to test this hypothesis.

The Chi-Square results are presented in Table 19. The observed relationship between the arable land per extension worker and farmer income was found to be not significant, χ^2 (4, N = 86) = 5.99, $p<.19$, at the .05 level of significance, therefore, the null hypothesis was not rejected. This result suggested that arable land per extension worker and farmer income are not related with each other. When this analysis was repeated only for the developing countries (per capita GDP of less than U.S. $ 5,000) the observed relationship continues to support the null hypothesis, χ^2 (4, N = 63) = 3.925, $p>.41$ at the .05 level of significance.

The results of multiplicative effect are presented in Table 20 indicated that fewer hectares of land (<550) per extension worker was related to increase farmer income of 44.14 per cent over the overall average ($e^{6.3686}$) 583.24. Having 550-2,000 hectares of arable land per extension worker was related to decrease farmer income of 16.22 per cent, and having >2,000 hectares of arable land per extension worker was related to decrease farmer income about 15 per cent.

Table 19: Frequency of Countries by Number of Hectares of Arable Land per Extension Worker and Farmer Income

Arable Land (ha)/ Extension Worker	*Farmer Income (US $)*			*Row Total*	*%*
	< 225	225-1,500	> 1,500		
< 550	7	8	12	27	31.4
550-2,000	10	12	5	27	31.4
> 2,000	14	8	10	32	37.2
Column Total	31	28	27	86	
(%)	36.0	32.6	31.4		100.00

Note— *N* Missing = 28, χ^2 = 5.996, *df* = 4, *p* = .1994, MEF = 8.477.

Table 20: Multiplicative Effect of Arable Land per Extension Worker on Farmer Income

Arable Land (ha)/ Extension Worker	*Mean*	*SD*	*Effect on Transformed Variables*	*Percentage Increase*	*P*
< 550	6.7342	1.5609	.3656	44.14	
550-2,000	6.1916	1.4910	–.1770	–16.22	} .6847
> 2,000	6.2059	3.4824	–.1627	–15.02	
	6.3686	2.5150			

Hypothesis 3: The two variables; hectares of arable land per extension worker and agricultural income/ha arable land, are independent of each other. Chi-Square analysis was used to test this hypothesis.

The Chi-Square results are presented in Table 21. The observed relationship between the arable land per extension worker and agricultural income/ha of arable land was found to be very significant, χ^2 (4, *N* = 86) = 23.534, *p*<.0001 at the .05 level of significance, therefore, the null hypothesis was not rejected. In practical terms, these findings indicate that as the per hectare area per extension worker decreases the agricultural income/ha arable land increases. When this analysis was repeated only for the developing countries (per capita GDP of less than

U.S. $ 5,000) the observed relationship between these two variables hold the same result, χ^2 (4, $N = 63$) = 22.792, $p <$ .001 at the .05 level of significance, hence, the null hypothesis was rejected.

Table 21: Frequency of Countries by Number of Hectares of Arable Land per Extension Worker and Agricultural Income

Arable Land (ha)/ Extension Worker	*Agricultural Income/ha (in US$)*			*Row Total*	*%*
	< 525	525-1,500	>1,500		
< 550	5	10	12	27	31.4
550-2,000	17	7	3	27	31.4
> 2,000	24	6	2	32	37.2
Column Total	46	23	17	86	
(%)	53.5	26.7	19.8		100.0

Note— *N* Missing = 28, χ^2 = 23.534, $df = 4$, p = .0001, MEF = 5.33.

The results of multiplicative effect are presented in Table 22 indicated that fewer hectares of arable land (<550) per extension worker was related to increase agricultural income of 291.92 per cent over the overall average ($e^{5.8543}$) 384.73. Having 550-2,000 hectares of arable land per extension worker increases agricultural income 0.37 per cent, and having >2,000 hectares of arable land per extension worker was related to decrease agricultural income about 69 per cent.

Table 22: Multiplicative Effect of Arable Land per Extension Worker on Agricultural Income

Arable Land (ha)/ Extension Worker	*Mean*	*SD*	*Effect on Transformed Variables*	*Percentage Increase*	*P*
< 550	7.2202	1.2561	1.3659	291.92	
550-2,000	5.8580	1.3958	0.0037	.37	} .0003
> 2,000	4.6736	4.6736	–1.1807	–69.29	
	5.8543	2.4035			

Hypothesis 4: The two variables; hectares of arable land per extension worker and total fertilizer consumption, are independent of each other. Chi-Square analysis was used to test this hypothesis.

The Chi-Square results are presented in Table 23. The observed relationship between the arable land per extension worker and fertilizer consumption kg/ha of arable land was found to be significant,

$\chi^2(4, N = 99) = 10.70, p<.05$, therefore, the null hypothesis was not rejected at the .05 level of significance. In practical terms, as the arable land per extension worker decreases, total fertilizer consumption increases. When this analysis was repeated only for the developing countries (per capita GDP of less than U.S. $ 5,000) the observed relationship between these two variables continue to hold the same relationship with $\chi^2 (4, N=78)=12.03, p.<.05$. Hence, the null hypothesis was rejected again at the .05 level of significance.

Table 23: Frequency of Countries by Number of Hectares of Arable Land per Extension Worker and Level of Total Fertilizer Consumption

Arable Land (ha)/ Extension Worker	*Total Fertilizer Consumption (in US$)*			*Row Total*	*%*
	< 25	25-100	>100		
< 550	7	8	16	31	31.3
550-2,000	13	9	9	31	31.3
> 2,000	15	16	6	37	37.4
Column Total	35	33	31	99	
(%)	35.4	33.3	31.3		100.0

Note. *N* Missing = 15, $\chi^2 = 10.705$, $df = 4$, $p = .030$, MEF = 9.707.

The results of multiplicative effect are presented in Table 24 indicated that fewer hectares of arable land (<550) per extension worker was related to decreases total fertilizer consumption kg/ha arable and permanent cropland 17.55 per cent over the overall average ($e^{2.0399}$) 7.69. Having 550-2,000 hectares of arable land per extension worker was related to increase fertilizer consumption 7.78 per cent and having >2,000 hectares of arable land per extension worker increases fertiliser consumption more than 10.96 per cent.

Table 24: Multiplicative Effect of Arable Land per Extension Worker on Total Fertilizer Consumption

Arable Land (ha)/ Extension Worker	*Mean*	*SD*	*Effect on Transforme d Variables*	*Percentage Increase*	*P*
< 550	1.8469	1.7213	–.1930	–17.55	
550-2,000	2.1148	1.8628	.0749	7.78	} .7686
> 2,000	2.1439	1.8900	.1040	10.96	
	2.0399	1.8155			

Agricultural Researchers

It is assumed that the number of agricultural researchers at work in each country is also directly linked with agricultural development through technological changes. Table 25 presents the average number of researchers by geographic regions and economic groups. An adequate number of researchers also depends on the size of the country and other factors, therefore, the number of farmers per researcher in different regions and economic groups were compared with different agricultural development variables. First, the average number of researchers are categorised by geographic regions and economic groups; these data are presented in Table 25. It is evident from this table that among the regions, Asia has the highest number (71,902) of agricultural researchers and Africa has the lowest (3,334) number (Figure 6). Among the economic groups, the low income country group has highest number of researchers (47,447, due largely to India and China) followed by high income (47,200), lower middle income (16,292) and upper middle income (8,327). To compare the intensity of Agricultural researchers, Table 26 presents the number of farmers per researcher.

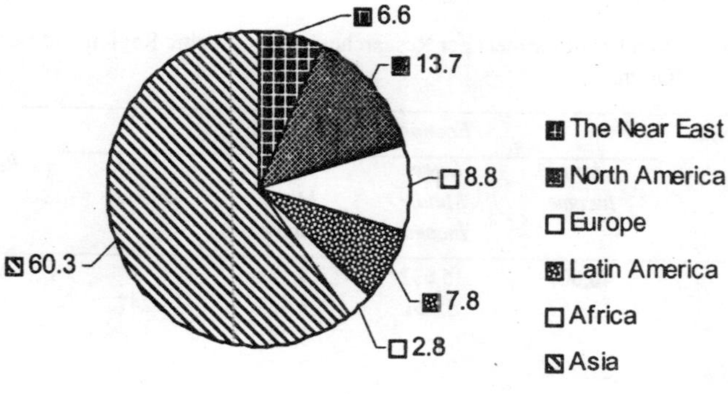

Fig. 6: Percentage of Researchers by Region

Number of Farmers per Agricultural Researcher

Next, it is important to have a general measure of agricultural research intensity. This measure is computed by determining the number of farmers per researcher. These findings are found in Table 26. This table indicates that North America has the lowest number of farmers per researcher, followed by Europe. The highest number of farmers per researcher were in Africa followed by Asia, Latin America, and the Near East (Figure 7). It is important to know if there is any association between the number of farmers per researcher with selected agricultural development indicators.

Table 25: Average Number of Researchers by Geographic Region and Economic Group

Geographic Region	*Economic Group (PCI)*				
	Low Income	*Lower Middle Income*	*Upper Middle Income*	*High Income*	*Region Total*
Africa	2,124	905	305	-	3,334 (31)
Asia	45,022	5,360	1,341	20,179	71,902 (26)
Europe	-	-	449	10,038	10,487 (14)
Latin America	88	3,005	6,232	-	9,324 (24)
North America	-	-	-	16,314	16,314 (3)
The Near East	213	7,022	-	669	7,904 (16)
Group Total	47,447 (37)	16,292 (41)	8,327 (12)	47,200 (24)	119,226 (114)

Note— Total number of countries is given in the parentheses.

Table 26: Number of Farmers per Researcher by Geographic Region and Economic Group

Geographic Region	*Economic Group (PCI)*				*Region*
	Low Income	*Lower Middle Income*	*Upper Middle Income*	*High Income*	
Africa	40,007	15,632	4,521	-	30,144
Asia	17,167	5,924	3,623	247	11,328
Europe	-	-	3,281	596	711
Latin America	21,739	7,160	2,438	-	4,143
North America	-	-	-	216	216
The Near East	22,235	2,760	2,840	3,291	
Group	18,221	5,328	2,751	347	8,309

Note— The number of farmers per researcher presented in this table was computed directly by dividing the total number of farmers given in Appendix D (Table D-3) by total number of researchers (Table 25) regardless of the number of countries. Therefore, the row and column totals of this table, do not add up to the total.

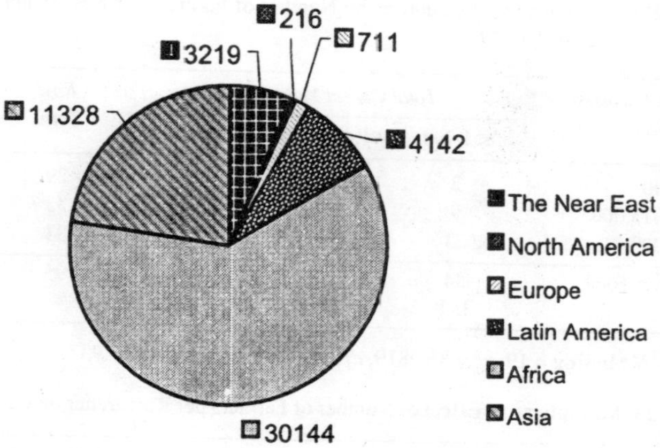

Fig. 7: Number of Farmers per Researcher by Region

Hence, the following four null hypotheses were tested to assess the potential relationship between the number of farmers per researcher in different countries with the four agricultural development indicators at the .05 level of significance.

Hypothesis 1: The variables, number of farmers per researcher and total cereal yield (kg/ha) are independent of each other.

The Chi-Square results are presented in Table 27. The observed relationship between the number of farmers per researcher and cereal yield was found to be very significant, $\chi^2(4, N = 95) = 35.98, p<.0001$ at the .05 level, therefore, the null hypothesis was rejected. It can be inferred from this result that fewer number of farmers per researcher is directly related with higher cereal yields. When this analysis was repeated only for the developing countries (per capita GDP of less than US $ 5,000), the observed relationship between these two variables continue to hold the same relationship χ^2 (4, $N = 77$) = 16.65, $p<.01$ level of significance.

The results of multiplicative effect are presented in Table 28 indicates that fewer number of farmers (<2,500) per researcher increases cereal yield (kg/ha) more than 62 per cent over the overall average ($e^{7.4597}$) 1736.6. Having 2,500-15,000 farmers per researcher decreases cereal yield more than 7 per cent, and having >15,000 farmers per researcher decreases total cereal yield more than 29 per cent.

Table 27: Frequency of Countries by Number of Farmers per Researcher and Total Cereal Yield

No. of Farmers/ Researcher	*Total Cereal Yield (kg/ha)*			*Row Total*	*%*
	< 1,400	1,400-2000	> 2,500		
< 2,500	2	11	16	29	30.6
2,500-15,000	9	18	6	33	34.7
> 15,000	23	6	4	33	34.7
Column Total	34	35	26	95	
(%)	35.8	36.8	27.4		100.0

Note— *N* Missing = 19, $\chi^2 = 35.9819$, $df = 4$, $p = .00000$, MEF = 7.937.

Table 28: Multiplicative Effect of Number of Farmers per Researcher on Cereal Yield

No. of Farmers/ Researcher	*Mean*	*SD*	*Effect on Transformed Variables*	*Percentage Increase*	*P*
< 2,500	7.9479	.4962	.4882	62.89	
2,500-15,000	7.3856	.5985	–.0741	–7.17	} .000
> 15,000	7.1047	.5304	–.355	–29.90	
	7.4597	.6409			

Hypothesis 2: The two variables; number of farmers per extension worker and farmer income, are independent of each other. Chi-Square analysis was used to test this hypothesis.

The Chi-Square results are presented in Table 29. The observed relationship between the number of farmers per researcher and farmer income was found to be very significant, $\chi^2(4, N = 84) = 38.71, p < .0001$ at the .05 level, therefore, the null hypothesis was rejected. It is evident from this table that fewer number of farmers per researcher were related to higher farmer income. When this analysis was repeated for only the developing countries (per capita GDP of less than US \$ 5,000), the observed relationship between these two variables continues to hold the same relationship $\chi^2(4, N = 61) = 23.56, p < .0001$ at .05 level of significance, therefore, the null hypothesis was rejected again.

The results of multiplicative effect are presented in Table 30 indicates that fewer number of farmers (<2,500) per researcher increases farmer income more than 478 per cent over the overall average ($e^{6.1865}$) 486.1. Having 2,500-15,000 farmers per researcher decreases farmer income more than 23 per cent, and having >15,000 farmers per researcher decreases farmer income more than 75 per cent.

Table 29: Frequency of Countries by Number of Farmers per Researcher and Farmer Income

No. of Farmers/ Researcher	*Farmer Income (US $)*			*Row Total*	*%*
	< 225	225-1,500	> 1,500		
< 2,500	2	7	17	26	31.33
2,500-15,000	8	15	5	28	33.37
> 15,000	21	6	2	29	34.94
Column Total	31	28	24	83	
(%)	37.3	33.7	28.9		100.0

Note— *N* Missing = 31, χ^2 = 38.711, *df* = 4, *p* = .00000, MEF = 7.429.

Table 30: Multiplicative Effect of Number of Farmers per Researcher on Farmer Income

No. of Farmers/ Researcher	*Mean*	*SD*	*Effect on Transformed Variables*	*Percentage Increase*	*P*
< 2,500	7.9456	2.5199	1.7591	478.69	
2,500-15,000	5.9208	2.4822	-.2657	.23.60	} .000
> 15,000	4.7939	1.5152	-1.3926	-75.24	
	6.1865	2.5435			

Hypothesis 3: The two variables; number of farmers per researcher and agricultural income/ha arable land are independent of each other. Chi-Square analysis was used to test this hypothesis.

The Chi-Square results are presented in Table 31. The observed relationship between the number of farmers per researcher and agricultural income/ha of arable land was found to be very significant, $\chi^2(4, N = 84) = 12.68, p < .05$ at the .05 level, therefore, the null hypothesis was rejected. Although the minimum expected frequency of one cell was 4.952, < 5 or (11.1%), the *p* value may be trusted. When this analysis was repeated for only the developing countries (per capita GDP of less than US $ 5,000), the observed relationship between these two variables does not hold the same relationship $\chi^2(4, N = 61) = 7.32, p > .11$. Hence, the null hypothesis that number of farmers per researcher and agricultural income/ha of arable land in developing countries are independent of each other could not be rejected. It should be pointed out however, that the χ^2 result was approaching significance, therefore, it is possible that there is a relationship between the two variables, but it was not detected at the .05 level of significance.

Table 31: Frequency of Countries by Number of Farmers per Researcher and Agricultural Income

No. of Farmers/ Researcher	*Agricultural Income (US $)*			*Row Total*	*%*
	< 525	525-1,500	>1,500		
< 2,500	8	8	10	26	31.0
2,500-15,000	19	6	4	29	34.5
> 15,000	20	7	2	29	34.5
Column Total	47	21	16	84	
(%)	56	25	19		100.0

Note— *N* Missing = 30, χ^2 = 12.682, df = 4, p = .01294, MEF = 4.952 cells with expected frequency <5- 1 of 9 (11.1%).

The results of multiplicative effect are presented in Table 32 indicates that fewer number of farmers (<2,500) per researcher increases agricultural income more than 85 per cent over the overall average ($e^{5.7183}$) 304.4. Having 2,500-15,000 farmers per researcher increases agricultural income more than 5 per cent, and having >15,000 farmers per researcher decreases agricultural income more than 46 per cent.

Table 32: Multiplicative Effect of Number of Farmers per Researcher on Agricultural Income

No. of Farmers/ Researcher	*Mean*	*SD*	*Effect on Transformed Variables*	*Percentage Increase*	*P*
< 2,500	6.3379	2.7155	.6196	85.50	
2,500-15,000	5.7756	2.7637	.0573	5.72	} .0000
> 15,000	5.0938	1.7279	–.6345	–46.54	
	5.7183	2.5413			

Hypothesis 4: The two variables; number of farmers per researcher and total fertilizer consumption are independent of each other. Chi-Square analysis was used to test this hypothesis.

The Chi-Square results are presented in Table 33. The observed relationship between the number of farmers per researcher and the fertilizer consumption (kg/ha) was found to be very significant, $\chi^2(4, N = 93) = 34.51, p<.0001$, therefore, the null hypothesis was rejected. This relationship indicates that the fewer number of farmers per researcher, is associated with higher level of fertilizer consumption. When this analysis was repeated for only the developing countries (per capita GDP of less than US $ 5,000), the observed relationship between these two variables continue to hold the same relationship, $\chi^2(4, N = 74) = 18.81, p > .001$. Hence, the null hypothesis was rejected again at the .05 level of significance.

Table 33: Frequency of Countries by Number of Farmers per Researcher and Total Fertilizer Consumption

No. of Farmers/ Researcher	*Total Fertilizer Consumption (kg/ha)*			*Row Total*	*%*
	< 25	25-100	> 100		
< 2,500	3	11	17	31	33.3
2,500-15,000	9	15	7	31	33.3
> 15,000	23	5	3	31	33.3
Column Total	35	31	27	93	
(%)	37.6	33.3	29.1		100.0

Note— *N* Missing = 21, $\chi^2 = 34.515$, $df = 4$, $p = .00000$, MEF = 9.000. The total of row percentage is rounded.

The results of multiplicative effect are presented in Table 34 indicates that fewer number of farmers (<2,500) per researcher increases total fertilizer consumption (kg/ha) more than 208 per cent over the overall average ($e^{1.9245}$) 6.85. Having 2,500-15,000 farmers per researcher increases total fertilizer consumption more than 10 per cent, and having >15,000 farmers per researcher decreases total fertilizer consumption more than 71 per cent.

Table 34: Multiplicative Effect of Number of Farmers per Researcher on Total Fertilizer Consumption

No. of Farmers/ Researchers	*Mean*	*SD*	*Effect on Transformed Variables*	*Percentage Increase*	*P*
< 2,500	3.0473	1.5125	1.1228	208.73	
2,500-15,000	2.0158	1.5362	0.0913	10.05	} .0000
> 15,000	.6743	.6953	−1.2502	−71.23	
	1.9245	1.8462			

Arable Land per Researcher

The number of hectares of arable land per researcher is another important measure of research intensity factor that may be related with agricultural development. Table 35 presents the number of hectares of arable land per researcher for the responding countries, grouped by geographic region and economic groups. The purpose of this table is to present the arable land (ha) per researcher by geographic region and economic groups. It is evident from the table that highest number of hectare of arable land per researcher (31,874) is found in Africa and the lowest in the Near East (5,273) (Figure 8). Among the economic groups, upper middle income group have highest hectares of arable land per researcher (13,261) and lowest was 6,727 hectare per researcher in the high income group. It is important to assess the relationship of arable land per researcher and agricultural development. The following four null hypotheses were tested to assess the relationship of arable land per researcher with four agricultural development indicators at the .05 level of significance.

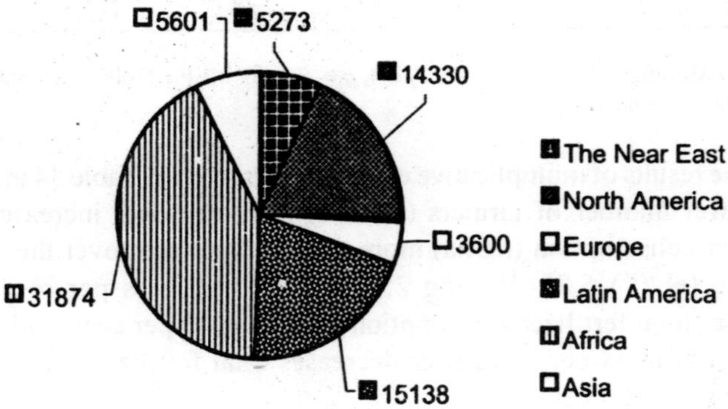

Fig. 8: Arable Land per Researcher by Geographic Region (ha)

Hypothesis 1: The variables, number of hectares of arable land per researcher and total cereal yield (kg/ha) are independent of each other. Chi-Square analysis was used to test this hypothesis.

The Chi-Square results are presented in Table 36. The observed relationship between the number of hectares of arable land per researcher and total cereal yield was found to be very significant χ^2 $(4, N = 97) = 39.14$, $p<.0001$ at the .05 level, therefore, the null hypothesis was rejected. When this analysis was repeated only for the developing countries (per capita GDP of less than U.S.$ 5,000), the observed relationship between these two variables continues to hold the same relationship, χ^2 (4, N = 79) = 15.11, p<.01, therefore, the null hypothesis was rejected again at the .05 level of significance.

Table 35: Arable Land (ha) per Researcher by Geographic Region and Economic Group

Geographic Region	*Economic Group (PCI)*				
	Low Income	*Lower Middle Income*	*Upper Middle Income*	*High Income*	*Region Total*
Africa	33,505	31,086	22,852	-	31,874
Asia	7,198	4,696	1,489	2,550	5,601
Europe	-	-	15,817	3,054	3,600
Latin America	11,761	15,226	15,141	-	15,138
North America	-	-	-	14,330	14,330
The Near East	42,455	4,418	-	2,404	5,273
Group Total	8,543	7,985	13,261	6,727	8,080

Note— The arable land per researcher presented in this table is computed directly by dividing the arable land in Appendix D (Table D-4) by average number of researcher (Table 25), regardless of the number of countries. Therefore, the row and column total of this table do not add up to the total.

Table 36: Frequency of Countries by Number of Hectares of Arable Land per Researcher and Total Cereal Yield

Arable Land (ha)/ Researcher	*Total Cereal Yield (kg/ha)*			*Row Total*	*%*
	< 1,400	1,400-2,500	> 2,500		
< 3,600	4	8	18	30	30.9
3,600-17,000	7	18	8	33	34.0
> 17,000	23	10	1	34	35.1
Column Total	34	36	27	97	
(%)	35.1	37.1	27.8		100.0

Note— N Missing = 17, χ^2 = 39.146, df = 4, p = .00000, MEF = 8.351.

The results of multiplicative effect are presented in Table 37, indicated that fewer hectares of land (<3,600) per researcher was related to increase total cereal yield (kg/ha) of 44.77 per cent over the overall average ($e^{7.4706}$) 1755.57. Having 3,600-17,000 hectares of arable land per researcher was related to increase cereal yield of 11.20 per cent, and having >17,000 hectares of arable land per researcher was related to decrease cereal yield of 34.72 per cent.

Table 37: Multiplicative Effect of Number of Hectares of Arable Land per Researcher on Cereal Yield

Arable Land (ha)/ Researcher	*Mean*	*SD*	*Effect on Transformed Variables*	*Percentage Increase*	*P*
< 3,600	7.8406	.7353	.3700	44.77	
3,600-17,000	7.5768	.4850	.1062	11.20	} .000
> 17,000	7.0410	.4336	–.4296	–34.72	
	7.4706	.6468			

Hypothesis 2: The variables, number of hectares of arable land per researcher and farmer income are independent of each other. Chi-Square analysis was used to test this hypothesis.

The Chi-Square results are presented in Table 38. The observed relationship between the number of hectares of arable land per researcher and farmer income was found to be very significant χ^2 (4, N = 85) = 33.41, p<.0001 at the .05 level, therefore, the null hypothesis was rejected. It is evident from the table that fewer hectares of arable land per researcher was related to increased levels of farmer income. When this analysis was repeated only for the developing countries (per capita GDP of less than U.S.$ 5,000), the observed relationship between these two variables holds the same relationship, χ^2 (4,N = 62) = 13.58, p<.01, therefore, the null hypothesis was rejected again at the .05 level of significance.

Table 38: Frequency of Countries by Number of Hectares of Arable Land per Researcher and Farmer Income

Arable Land (ha)/ Researcher	*Farmer Income (US$)*			*Row Total*	*%*
	< 225	225-1,500	> 1,500		
< 3,600	2	6	16	24	28.2
3,600-17,000	10	14	8	32	37.7
> 17,000	20	8	1	29	34.1
Column Total	32	28	25	85	
(%)	37.7	32.9	29.4		100.00

Note— N Missing = 29, χ^2 = 33.412, df = 4, p = .0000, MEF = 7.059.

The results of multiplicative effect are presented in Table 39 indicated that fewer hectares of land (<3,600) per researcher was related to increase farmer income of 380.04 per cent over the overall average ($e^{6.2093}$) 497.35.

Having 3,600-17,000 hectares of arable land per researcher was related to increase farmer income of 41 per cent, and having >17,000 hectares of arable land per researcher was related to decrease farmer income about 81per cent.

Table 39: Multiplicative Effect of Number of Hectares of Arable Land per Researcher on Farmer Income

Arable Land (ha)/ Researcher	*Mean*	*SD*	*Effect on Transformed Variables*	*Percentage Increase*	*P*
< 3,600	7.7780	1.5117	1.5687	380.04	
3,600-17,000	6.5529	1.8416	.3436	41.00	}.0000
> 17,000	4.5457	2.9464	–1.6636	–81.05	
	6.2093	2.5346			

Hypothesis 3: The variables, number of hectares of arable land per researcher and agricultural income per hectare of arable land are independent of each other. Chi-Square analysis was used to test this hypothesis.

The Chi-Square results are presented in Table 40. The observed relationship between the number of hectares of arable land per researcher and agricultural income was found to be very significant χ^2 (4, N = 85) = 52.03, p<.0001, therefore, the null hypothesis was rejected. Although one cell count was 0 (minimum expected frequency was 4.8) in the table, the percentage of cells with expected frequency was however, within the acceptable limits (11.1 per cent) of less than 25 per cent. Hence, the p may be trusted, but not the χ^2 value. When this analysis was repeated only for the developing countries (per capita GDP of less than U.S.$ 5,000), the observed relationship between these two variables holds the same relationship, χ^2 (4, N = 62) = 23.73, p<.0001 at the .05 level of significance. Once cell in this set was also have 0 count but minimum expected frequency was > 5, (5.5), therefore, the p may be trusted, but not the χ^2 value.

The results of multiplicative effect are presented in Table 41 indicated that fewer hectares of land (<3,600) per researcher was related to increase agricultural income of 539.43 per cent over the overall average ($e^{5.7518}$) 314.76. Having 3,600-17,000 hectares of arable land per researcher was related to increase agricultural income of 22.19 per cent, and having >17,000 hectares of arable land per researcher was related to decrease agricultural income about 81 per cent.

Table 40: Frequency of Countries by the Number of Hectares of Arable Land per Researcher and Agricultural Income

Arable Land (ha)/ Researcher	*Agricultural Income/ha (US $)*			*Row Total*	*%*
	< 525	525-1,500	> 1,500		
< 3,600	1	8	15	24	28.2
3,600-17,000	20	10	2	32	37.7
> 17,000	26	3	0	29	34.1
Column Total	47	21	17	85	
(%)	55.3	24.7	20.0		100.00

Note— *N* Missing = 29, $\chi^2 = 52.037$, $df = 4$, $p = .00000$, MEF = 4.8, cells with expected frequency <5-1 of 9 (11.1%).

Table 41: Multiplicative Effect of the Hectares of Arable Land (ha) per Researcher on Agricultural Income

Arable Land (ha)/ Researcher	*Mean*	*SD*	*Effect on Transformed Variables*	*Percentage Increase*	*P*
< 3,600	7.6072	.9251	1.8554	539.43	
3,600-17,000	5.9522	1.2848	.2004	22.19	}.0000
> 17,000	4.0220	3.1043	–1.7298	–81.05	
	5.7518	2.4527			

Hypothesis 4: The variables, number of hectares of arable land per researcher and total fertilizer consumption (kg/ha) are independent of each other. Chi-Square analysis was used to test this hypothesis.

The Chi-Square results are presented in Table 42. The observed relationship between the number of hectares of arable land per researcher and total fertilizer consumption (kg/ha) was found to be very significant χ^2 (4, N = 94) = 56.11, p<.0001, therefore, the null hypothesis was rejected. Results indicated that arable land per researcher was related to higher level of fertilizer consumption. Although one cell count was 0 the minimum expected frequency was however more than 5 (8.638). Hence, the p can be trusted, but not the χ^2. When this analysis was repeated only for the developing countries (per capita GDP of less than U.S.$ 5,000), the observed relationship between these two variables holds the same relationship, χ^2 (4, N = 75) = 28.41, p<.0001, hence the null hypothesis was rejected again at the .05 level of significance.

The results of multiplicative effect are presented in Table 43 indicated that fewer hectares of land (<3,600) per researcher was related to increase total fertilizer consumption kg/ha arable land and permanent cropland of 58.93 per cent over the overall average ($e^{1.9426}$) 6.98. Having 3,600-17,000 hectares of arable land per researcher was related to increase fertilizer consumption of 74.8 per cent, and having >17,000 hectares of arable land per researcher was related to decrease fertilizer consumption more than 63 per cent.

Table 42: Frequency of Countries by the Number of Hectares of Arable Land per Researcher and Total Fertilizer Consumption (kg/ha)

Arable Land (ha)/ Researcher	*Total Fertilizer Consumption*			*Row Total*	*%*
	<25	25-100	>100		
< 3,600	4	5	20	29	30.9
3,600-17,000	6	19	8	33	35.1
> 17,000	25	7	0	32	34.0
Column Total	35	31	28	94	
(%)	37.2	33.0	29.8		100.0

Note. *N* Missing = 20, χ^2 = 56.111, *df* = 4, *p* = .00000, MEF = 8.638.

Table 43: Multiplicative Effect of Number of Hectares of Arable Land per Researcher on Fertilizer Consumption

Arable Land (ha)/ Researcher	*Mean*	*SD*	*Effect on Transformed Variables*	*Percentage Increase*	*P*
< 3,600	2.4059	1.9553	.4633	58.93	
3,600-17,000	2.5068	1.2990	.5642	57.80	}.0004
> 17,000	2.9262	1.8406	−1.0164	−63.81	
	1.9426	1.8447			

Extension Personnel and Researcher

The number of extension personnel per researcher is another indicator showing the allocation of human resources within an agricultural technology system. Data on the number of extension workers per researcher, for 94 countries are presented in Table 44. This table indicates the number of extension workers/researcher in Africa is very high 18/1, where as it is very low in Europe and North America 1/1 (Figure 9). Among the economic groups, the lower income groups have a higher proportion of extension workers/researcher in comparison to higher income groups. The worldwide average is 5/1. It is therefore, important to find out if this ratio is related with the agricultural development indicators.

Table 44: Number of Extension Personnel per Researcher by Geographic Region and Economic Group

Geographic Region	*Economic Group (PCI)*				*Region*
	Low Income	*Lower Middle Income*	*Upper Middle Income*	*High Income*	
Africa	21	16	4	-	18
Asia	7	8	6	1	5
Europe	-	-	5	1	1
Latin America	12	9	.3	-	3
North America	-	-	-	1	1
The Near East	5	4	2	4	
Group	8	7	2	1	5

Note— The number of extension personnel per researcher that are presented in this table was computed directly by dividing the total number of extension personnel given in Table 6 by average number of researchers presented in Table 25, regardless of the number of countries. Therefore, the row and column totals of this table, do not add up to the total. Numbers of more than .5 were rounded up to 1 unit in this table. $N = 94$, Missing = 20.

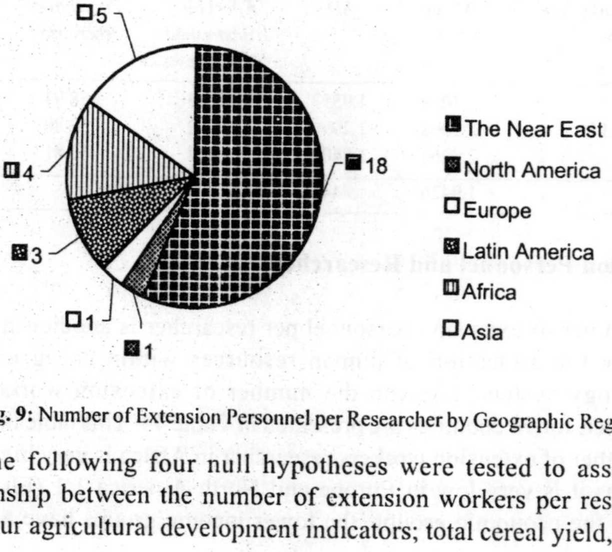

Fig. 9: Number of Extension Personnel per Researcher by Geographic Region

The following four null hypotheses were tested to assess the relationship between the number of extension workers per researcher with four agricultural development indicators; total cereal yield, farmer income, agricultural income/hectare of arable land and total fertiliser consumption in kg/ha of arable and permanent cropland, at .05 level of significance.

Hypothesis 1: The variables, number of extension workers per researcher and total cereal yield are independent of each other. Chi-Square analysis was used to test this hypothesis.

The Chi-Square results are presented in Table 45. The observed relationship between the number of extension personnel per researcher and total cereal yield (kg/ha) was found to be significant, χ^2 (4, N = 88) = 10.145, $p<.05$, therefore, the null hypothesis was rejected. This finding suggests that as the relative proportion of researchers to extension workers increases that cereal yields also tend to increase. When this analysis was repeated only for the developing countries (per capita GDP of less than US $ 5000), the observed relationship between the number of extension personnel per researcher and total cereal yield does not hold the same relationship, χ^2 (4, N = 75) = 4.606, $p>.33$, hence, the null hypothesis was rejected at .05 level of significance.

Table 45: Frequency of Countries by Number of Extension Worker per Researcher and Total Cereals Yield

No. of Extension Worker/Researcher	*Total Cereal Yield (kg/ha)*			*Row Total*	*%*
	< 1,400	1,400-2,500	> 2,500		
< 5	7	13	13	33	37.5
5-10	6	10	6	22	25.0
> 10	18	10	5	33	37.5
Column Total	31	33	24	88	
(%)	35.2	37.5	27.3		100.0

Note— *N* Missing = 26, χ^2 = 10.1453, df = 4, p = .0380, MEF = 8.74.

The results of multiplicative effect presented in Table 46 indicates that fewer number of extension workers (<5) per researcher decreases total cereals yield kg/ha 9.31 per cent over the overall average ($e^{7.4575}$) 1728.8. Having 5-10 extension worker per researcher decreases total cereal yield 21.09 per cent, and having >10 extension worker per researcher decreases more than 40 per cent total cereal yield.

Hypothesis 2: The variables, extension worker per researcher and farmer income are independent of each other. Chi-Square analysis was used to test this hypothesis.

The Chi-Square results are presented in Table 47. The observed relationship between extension personnel per researcher and farmer income was found to be significant, χ^2 (4, N = 79) = 13.098, p<.05, therefore, the null hypothesis was rejected. It can be inferred that higher farmer income is related to a more balanced ratio of extension workers to researcher (*i.e.* < 5 EW/R). When this analysis was repeated for the developing countries (per capita GDP of less than US $ 5,000), the observed relationship continues to hold the relationship, χ^2 (4, N = 65) = 21.667, p<.001 at the .05 level of significance.

Table 46: Multiplicative Effect of Number of Extension Worker per Researcher on Cereal Yield

No. of Extension Worker/ Researcher	*Mean*	*SD*	*Effect on Transformed Variables*	*Percentage Increase*	*P*
< 5	7.6498	.6898	.1923	–9.31	
5-10	7.5107	.5363	.0532	–21.09	}.0218
> 10	7.2297	.5734	–.2278	–40.42	
	7.4575	.6323			

Table 47: Frequency of Countries by Number of Extension Worker per Researcher and Farmer Income

No. of Extension Worker per Researcher	*Farmer Income (US $)*			*Row Total*	*%*
	< 225	225-1,500	> 1,500		
< 5	7	9	13	29	36.7
5-10	7	10	4	21	26.6
> 10	17	9	3	29	36.7
Column Total	31	28	20	79	
(%)	39.3	35.4	25.3		100.00

Note— N Missing = 35, χ^2 = 13.098, df = 4, p = .0108, MEF = 5.316.

The results of multiplicative effect presented in Table 48 indicates that fewer number of extension worker (<5) per researcher increases farmer income 88.161 per cent over the overall average ($e^{6.0717}$) 433.42. Having 5-10 extension worker per researcher increases farmer income by 5.88 per cent, and having >10 extension worker per researcher decreases farmer income more than 55 per cent.

Hypothesis 3: The variables, number of extension worker per researcher and agricultural income, are independent of one another. Chi-Square analysis was used to test this hypothesis.

Table 48: Multiplicative Effect of Number of Extension Worker per Researcher on Farmer Income

No. of Extension Worker/Researcher	*Mean*	*SD*	*Effect on Transformed Variables*	*Percentage Increase*	*P*
< 5	6.7038	3.2373	.6321	88.16	
5-10	6.1288	1.0318	.0571	5.88	}.0967
> 10	5.2629	1.8378	–.8088	–55.46	
	6.0717	2.4253			

The Chi-Square results are presented in Table 49. The observed relationship between number of extension worker per researcher and agricultural income per hectare arable land was not significant, χ^2 (4, N = 79) = 6.509, p<.16 at the .05 level of significance, therefore the null hypothesis was accepted. However, since the level of significance was approaching the .05 level, there may be a relationship between these variables, but it was not detected. Minimum expected frequency of one cell out of 9 cells was also < 5 (3.987), but it is only 11.1% which is within the acceptable limit (25%). The p value therefore may be trusted, but not the χ^2. When this analysis was repeated only for the developing countries where (per capita GDP of less than US $ 5,000), the observed relationship between these two variables holds the same relationship χ^2 (4, N = 61) = 7.096, p>.13 at the .05 level of significance, hence, the null hypothesis was accepted.

Table 49: Frequency of Countries by the Number of Extension Worker per Researcher and Agricultural Income

No. of Extension Worker/Researcher	*Agricultural Income /ha (US $)*			*Row Total*	*%*
	< 525	525-1.500	> 1,500		
< 5	11	10	8	29	36.7
5-10	13	4	4	21	26.6
> 10	20	6	3	29	36.7
Column Total	44	20	15	79	
(%)	55.7	25.3	19.0		100.00

Note— N Missing = 35, χ^2 = 6.509, df = 4, p = .1642, MEF = 3.987.

The results of multiplicative effect presented in Table 50 indicates that fewer number of extension workers (<5) per researcher increases agricultural income 6.4 per cent over the overall average ($e^{5.761}$) 317.67. Having 5-10 extension workers per researcher increases agricultural income 59.19 per cent, and having >10 extension worker per researcher decreases agricultural income more than 35 per cent.

Table 50: Multiplicative Effect of Number of Extension Worker per Researcher on Agricultural Income

No. of Extension Worker per Researcher	*Mean*	*SD*	*Effect on Transformed Variables*	*Percentage Increase*	*P*
< 5	5.8230	3.4180	.0620	6.40	
5-10	6.2259	1.1815	.4649	59.19	}.4887
> 10	5.3180	1.7880	–.443	–35.79	
	5.7610	2.4744			

Hypothesis 4: The variables, number of extension workers per researcher and total fertilizer consumption in kg/ha arable and permanent cropland are independent of each other.

The Chi-Square results are presented in Table 51. The observed relationship between the extension worker per researcher and total fertilizer consumption was found to be very significant, χ^2 (4, N = 85) = 13.368, p<.01, therefore, the null hypothesis was rejected. It is evident from these findings that fewer number of extension workers per researcher results in increasing levels of fertilizer consumption. When this analysis was repeated for only the developing countries (per capita GDP of less than US $ 5,000), the observed relationship between these two variables hold the same relationship, χ^2 (4, N = 71) = 13.953, p<.01, at the .05 level of significance.

Table 51: Frequency of Countries by the Number of Extension Worker per Researcher and Total Fertilizer Consumption

No. of Extension Worker/Researcher	*Total Fertilizer Consumption (kg/ha)*			*Row Total*	*%*
	< 25	25-100	> 100		
< 5	7	10	17	34	38.8
5-10	8	8	6	22	35.9
> 10	17	10	3	30	35.3
Column Total	32	28	26	86	
(%)	37.7	32.9	29.4		100.00

Note— *N* Missing = 29, χ^2 = 13.368, *df* = 4, *p* = .00961, MEF = 6.471.

The results of multiplicative effect presented in Table 52 indicate that fewer number of extension worker (<5) per researcher increases total fertilizer consumption (kg/ha) arable and permanent cropland more than 95 per cent over the overall average ($e^{1.9504}$) 7.03. Having 5-10 extension worker per researcher increases total fertilizer consumption 18.10 per cent, and having >10 extension worker per researcher decreases fertilizer consumption more than 58 per cent.

Table 52: Multiplicative Effect of Number of Extension Worker per Researcher on Fertilizer Consumption

No. of Extension Worker per Researcher	*Mean*	*SD*	*Effect on Transformed Variables*	*Percentage Increase*	*P*
<5	2.6185	1.5951	.6681	95.05	
5-10	2.1168	1.1844	.1664	18.10	} .0006
>10	1.0713	1.7246	–.8791	–58.48	
	1.9504	1.6756			

Subject-Matter Specialists

Subject-matter specialists (SMSs) play a critical role in an extension system. They provide much of the linkage between the research and extension system and provide technical backstoping and in-service technical training for the extension field staff. The success of an extension system in technology transfer is directly influenced by the proportion and quality of SMSs. Based on other research (see Swanson *et al.*, 1990), it appears that a functionally effective SMS to agent ratio would be approximately 1 to 4-5, or about 20 per cent SMSs with at least M. S. degree-level or equivalent training and extensive field experience. Data on the percentage of SMSs in each national extension system is presented by geographic regions and economic groups in Table 53.

It is evident from this table that the percentage of subject-matter specialists is considerably low in Latin America, Africa, and the Near East in comparison to North America and Europe (Figure 10); the same type of situation is found in low income and lower middle income countries. It is, therefore, important to test and determine if the percentage of SMS is related to the agricultural development level in different countries. The following four null hypotheses were used to assess the possible relationship between the percentage of SMS of total extension personnel, with the four agricultural development indicators; total cereal yield, farmer income, agricultural income, and fertilizer consumption in kg/ha of arable and permanent cropland, at the .05 level of significance.

Table 53: Percentage of SMS of Total Extension Personnel by Geographic Region and Economic Groups (Mean)

Geographic Regions	*Economic Group (PCI*)*				*Region Mean*
	Low Income	*Lower Middle Income*	*Upper Middle Income*	*High Income*	
Africa	12.5 (20)	12.3 (7)	15.4 (1)	-	12.5 (28)
Asia	5.5 (10)	12.1 (10)	40.2 (1)	34.2 (3)	13.3 (24)
Europe	-	-	13.4 (2)	25.5 (7)	22.8 (9)
Latin America	5.8 (2)	13.2 (15)	10.6 (7)	-	11.6 (24)
North America	-	-	-	45.2 (3)	45.2 (3)
The Near East	10.6 (3)	18.8 (5)	-	9.7 (6)	13.1 (14)
Group Mean	9.95 (35)	13.49 (37)	13.79 (10)	25.0 (19)	14.4 (102)

* PCI = Per Capita Income

Note— Number of countries is given in the parentheses. $N = 102$, Missing = 12.

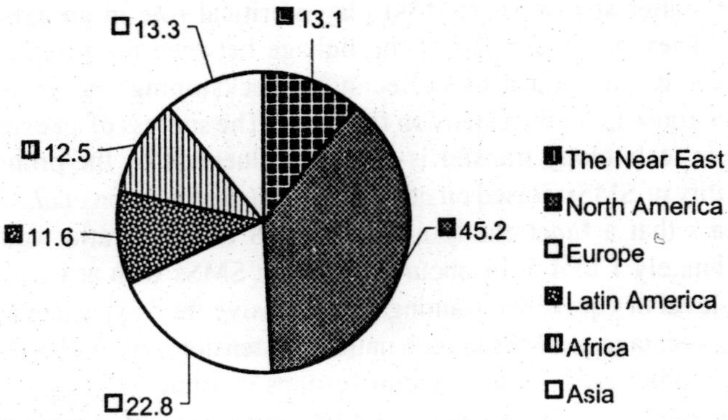

Fig. 10: Percentage of SMS of Total Extension Personnel by Geographic Region

Hypothesis 1: The variables, percentage of SMS and total cereal yield (kg/ha) are independent of each other. Chi-Square analysis was used to test this hypothesis.

The Chi-Square results are presented in Table 54. The observed relationship between the percentage of SMS and total cereal yield was not found to be significant χ^2 (4, N = 95) = 4.07, p>.31 at the .05 level of significance. Therefore, the null hypothesis could not be rejected. When this analysis was repeated for only the developing countries (per capita GDP of less than U.S. $ 5,000) included in the data set, the same relationship was observed, χ^2 (4, N = 79) = 7.75, p<.10, at the .05 level of significance.

Table 54: Frequency of Countries by the Percentage of SMS and Total Cereal Yield (kg/ha)

Percentage of SMS in Extension System	*Total Cereal Yield (kg/ha)* < 1400	1400-2500	> 2500	*Row Total*	*%*
<7	9	17	8	34	35.8
7-15	13	7	9	29	30.5
> 15	10	13	9	32	33.7
Column Total	32	37	26	95	
(%)	33.7	38.9	27.4		100.0

Note— *N* Missing = 19, χ^2 = 4.705, df = 4, p = .3188, MEF = 7.937

The results of multiplicative effect are presented in Table 55, indicate that fewer percentage of SMS (< 7 per cent) of total extension worker indicated decrease of 0.8 per cent total cereal yield (kg/ha) over the overall average ($e^{7.4804}$)1772.95. Having 7-15 per cent of SMS of total extension personnel indicated decrease of 4.17 per cent cereal yield. But having >15 per cent SMS of total extension personnel indicated increase of 4 per cent total cereal yield.

Table 55: Multiplicative Effect of Percentage of SMS on Cereal Yield

Percentage of SMS in Extension System	*Mean*	*SD*	*Effect on Transformed Variables*	*Percentage Increase*	*P*
< 7	7.4796	.5254	–.0008	–0.08	
7-15	7.4378	.6956	–.0426	–4.17	}.8861
> 15	7.5198	.7250	.0394	4.02	
	7.4804	.6446			

Hypothesis 2: The variables, percentage of SMS and farmer income are independent of each other.

The Chi-Square results are presented in Table 56. The observed relationship between the percentage of SMS and farmer income was not found to be significant χ^2 (4, N = 83) = 5.74, p<.21, at the .05 level of significance. Therefore, the null hypothesis could not be rejected. When this analysis was repeated only for the developing countries (per capita GDP of less than U.S. $ 5,000), the observed relationship between these two variables holds the same relationship, χ^2 (4, N = 61) = .70, p<.95, at the .05 level of significance.

Table 56: Frequency of Countries by the Percentage of SMS and Farmer Income

Percentage of SMS in Extension System	*Farmer Income (US $)* < 225	225-1500	> 1500	*Row Total*	*%*
< 7	15	5	10	30	36.2
7-15	9	11	6	26	31.3
> 15	8	9	10	27	32.5
Column Total	32	25	26	83	
(%)	38.6	30.1	31.3		100.0

Note— *N* Missing = 31, χ^2 = 5.743, df = 4, p = .2191, MEF = 7.831.

The results of multiplicative effect are presented in Table 57, indicate that fewer percentage of SMS (< 7 per cent) of total extension worker decreases former income 46.94 per cent over the overall average ($e^{6.2527}$) 519.41. Having 7-15 per cent of SMS of total extension personnel indicated increase of 7.98 per cent in farmer income and having >15 per cent SMS indicated increase of more than 70 per cent in farmer income. In other words more the percentage of SMSs of total extension workers more the farmer income.

Table 57: Multiplicative Effect of Percentage of SMS on Farmer Income

Percentage of SMS in Extension System	*Mean*	*SD*	*Effect on Transformed Variables*	*Percentage Increase*	*P*
< 7	5.6190	2.7393	-0.6337	-46.94	
7-15	6.3295	1.7697	0.0768	7.98	}.2701
> 15	6.7881	3.0191	0.5354	70.81	
	6.2527	2.5840			

Hypothesis 3. The variables, percentage of subject-matter specialists and agricultural income/ha of arable land are independent of each other. Chi-Square analysis was used to test this hypothesis.

The Chi-Square results are presented in Table 58. The observed relationship between percentage of SMS and agricultural income/ha of arable land was found to be not significant $\chi^2 (4, N = 83) = 7.283, p < .12$, at the .05 level of significance. Therefore, the null hypothesis could not be rejected. It appears that there is no direct relationship between these variables. When this analysis was repeated only for the developing countries (per capita GDP of less than U.S. $ 5,000), the observed relationship between these two variables holds the same relationship, $\chi^2 (4, N = 61) = 6.244, p > .18$, at the .05 level of significance. Hence the null hypothesis was accepted.

Table 58: Frequency of Countries by the Percentage of SMS and Agricultural Income of Arable Land

Percentage of SMS in Extension System	*Agricultural Income (US $)*			*Row Total*	*%*
	<525	525-1500	>1500		
< 7	18	5	7	30	36.2
7-15	10	12	4	26	31.3
> 15	17	6	4	27	32.5
Column Total	45	23	15	83	
(%)	54.2	27.7	18.1		100.0

The results of multiplicative effect are presented in Table 59, indicate that fewer percentage of SMS (< 7 per cent) of total extension worker decreases agricultural income 36.76 per cent over the overall average ($e^{5.7596}$) 317.22. Having 7-15 per cent of SMS of total extension personnel indicated 85.52 per cent increase in agricultural income and having > 15 per cent SMS indicated more than 14 per cent decrease in agricultural income. It may be inferred that less than 7 per cent and more than 15 per cent of SMS are indicating negative affect but 7-15 per cent of SMS was appropriate for increasing agricultural income.

Hypothesis 4: The variables, percentage of SMS and total fertilizer consumption (kg/ha) are independent of each other. Chi-Square analysis was used to test this hypothesis.

The Chi-Square results are presented in Table 60. The observed relationship between percentage of SMS and total fertilizer consumption was not found to be significant $\chi^2 (4, N = 93) = 2.61, p > .62$, at the .05

level of significance. Therefore, the null hypothesis could not be rejected. It appears that there is no relationship between these two variables. When this analysis was repeated only for the developing countries (per capita GDP of less than U.S. $ 5,000), the observed relationship between these two variables holds the same relationship, χ^2 (4, N = 75) = 1.844, p>.76, at the .05 level of significance. Hence the null hypothesis was accepted.

Table 59: Multiplicative Effect of Percentage of SMS on Agricultural Income

Percentage of SMS in Extension System	*Mean*	*SD*	*Effect on Transformed Variables*	*Percentage Increase*	*p*
< 7	5.3013	2.9911	–0.4583	–36.76	
7-15	6.3776	1.3990	0.6180	85.52	}.2863
> 15	5.5060	2.6891	–0.1536	–14.24	
	5.7596	2.4649			

Table 60: Frequency of Countries by the Percentage of SMS and Total Fertilizer Consumption (kg/ha)

Percentage of SMS in Extension System	*Total Fertilizer Consumption*			*Row Total*	*%*
	< 25	25-100	> 100		
< 7	14	10	9	33	35.5
7-15	11	9	6	26	28.0
> 15	9	13	12	34	36.5
Column Total	34	32	27	93	
(%)	36.6	34.4	29.0		100.0

Note—*N* Missing = 21, χ^2 = 2.610, df = 4, p = .6249, MEF = 7.548.

The results of multiplicative effect are presented in Table 61, indicated that fewer percentage of SMS (< 7 per cent) of total extension worker decreases 24.82 per cent total fertilizer consumption (kg/ha arable and permanent cropland) over the overall average ($e^{1.9452}$) 6.99. Having 7-15 per cent of SMS of total extension personnel indicated 33.76 per cent decrease in fertilizer consumption and having >15 per cent SMS indicated more than 77 per cent increase in fertilizer consumption.

Female Extension Personnel

It has been expressed by the extension scholars and other policy analysts that female farmers are bypassed by the male extension workers. It is widely accepted that female extension personnel will be more successful in reaching female farmers with advisory services. Female extension personnel are more likely to be influential and effective in reaching female farmers, than male extension workers will be, regardless of socio-cultural, agro-climatic and politico-economic conditions. But the fact is that the proportion of female extension personnel is very low. Hence, more female may be needed as extension administrators, subject-matter specialists, and field staff to help reach and educate female farmers. In order to justify the demand of more female extension personnel, the empirical support and evidence will be more convincing to the Government, Donor agencies, and national extension organisations worldwide.

Table 61: Multiplicative Effect of Percentage of SMS on Total Fertilizer Consumption

Percentage of SMS in Extension System	*Mean*	*SD*	*Effect on Transformed Variables*	*Percentage Increase*	*p*
< 7	1.6599	1.7001	–0.2853	–24.82	
7-15	1.5333	1.9653	–0.4119	–33.76	}.0685
> 15	2.5202	2.8673	0.5750	77.71	
	1.9452	1.8731			

The decision to increase the female extension personnel, therefore, depends on two factors. The first factor is demand for female extension personnel. The second factor may be determined by looking at the percentage of female population who are economically active in agriculture. If the population of the female population who are economically active in agriculture is higher, then it is assumed that the demand for female extension personnel may also be higher; hence, the proportion of female extension personnel needs to increase.

The second criterion is to test whether or not the proportion of female extension personnel is significantly related with the agricultural development indicators. If the proportion of female extension personnel is found to be empirically related to the agricultural development process, then there is supporting evidence to support the policy initiative that the number of female extension personnel should be increased. Main focus of this part is therefore, to analyse and present the estimated percentage of the females that are economically active in agriculture and then analyse the percentage of female extension personnel. Finally, Chi-Square analysis is performed to determine whether or not the percentage of female extension personnel is related with the agricultural development indicator (Cereal Yield, farmer income, Agricultural income and fertilizer consumption).

To estimate the demand criteria for female extension personnel, Table 62 presents the percentage of economically active female population in agriculture by geographic regions and economic groups. It is evident from this table that women farmers are comparatively higher in Africa (39%) as contrasted with the rest of the regions (Figure 11). But the situation of female extension personnel is just reverse (see Table 63) (Figure 12). Overall, more than 26 per cent of all farmers worldwide are females but the percentage of female extension workers is barely 11 per cent. This finding suggests that there are inadequate female extension personnel in most responding countries.

Table 62: Estimated Percentage of Females in Economically Active Population in Agriculture by Geographic Region and Economic Group (Mean 1985)

Geographic Region	*Economic Group (PCI)*				
	Low Income	*Lower Middle Income*	*Upper Middle Income*	*High Income*	*Region Total*
Africa	43.1 (21)	33.4 (9)	2.6 (1)	—	39.0 (31)
Asia	31.0 (11)	25.1 (9)	46.2 (1)	32.4 (3)	24.0 (24)
Europe	—	—	35.4 (3)	26.8 (11)	28.6(14)
Latin America	22.0 (2)	9.3 (16)	19.6 (5)	—	12.6 (23)
North America	—	—	—	16.2 (3)	16.2 (3)
The Near East	20.5 (3)	27.5 (6)	—	10.9 (7)	18.9(16)
Group Mean	36.56 (37)	21.04 (40)	23.26 (11)	22.35 (23)	26.7 (111)

Note— Number of countries is given in the parentheses. Data of estimated percentage of female in economically active population in agriculture were taken from the worldwide estimates and projections of the Agricultural and Non-agricultural population segments, 1950-2025, FAO, 1986.

The second criterion is to determine if a higher proportion of female extension personnel is associated with increased levels of agricultural development. In order to test it the following four null hypotheses were be tested to assess the relationship between the percentage of female extension personnel with four agricultural development indicators; total cereal yield, farmer income, agricultural income/ha of arable land, and total fertilizer consumption in kg/ha of arable and permanent cropland, at the .05 level of significance.

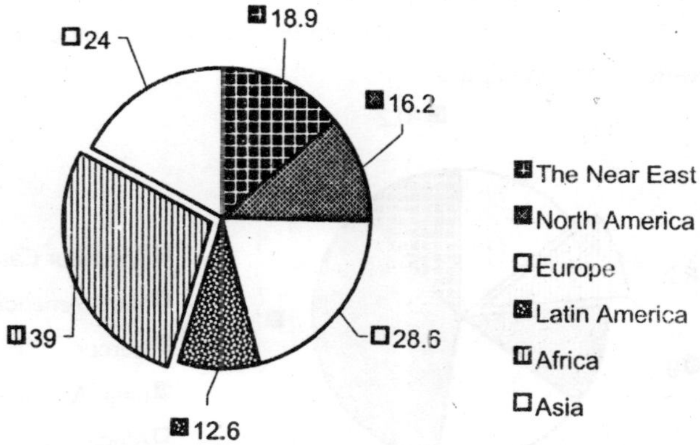

Fig. 11: Percentage of Females in Economically Active Population in Agriculture

Table 63: Percentage of Female Extension Personnel by Geographic Region and Economic Group (Mean)

Geographic Region	*Economic Group (PCI)*				
	Low Income	*Lower Middle Income*	*Upper Middle Income*	*High Income*	*Region Total*
Africa	9.6 (19)	4.4 (7)	—	—	8.2 (26)
Asia	6.6 (8)	18.1 (7)	4.5 (1)	15.1 (3)	12.1 (19)
Europe	—	—	18.9 (2)	13.9 (6)	15.1 (8)
Latin America	8.5 (2)	10.0 (14)	10.0 (3)	—	9.9 (21)
North America	—	—	—	30.7 (3)	30.7 (3)
The Near East	5.3 (3)	15.4 (5)	—	4.7 (2)	10.2(10)
Group Mean	8.4 (32)	11.4 (33)	11.5 (8)	16.4 (14)	11.1 (87)

Note—Number of countries is given in the parentheses, $N = 87$, Missing = 27.

Hypothesis 1: The variables, percentage of female extension workers and total cereal yield are independent of each other. Chi-Square analysis was used to test this hypothesis.

The Chi-Square results are presented in Table 64. The observed relationship between percentage of female extension personnel and total cereal yield was found to be not significant, $\chi^2(4, N = 82) = 7.722, p > .10$, at the .05 level of significance, therefore, the null hypothesis was accepted. When this analysis was repeated only for the developing countries (per capita GDP of less than U.S. $ 5,000), the observed relationship between these two variables holds the same relationship, $\chi^2(4, N = 70) = 1.567, p > .81$, at the .05 level of significance, hence the null hypothesis was accepted.

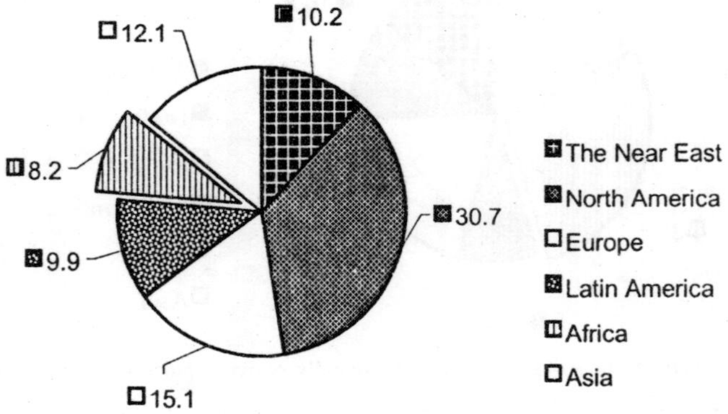

Fig. 12 Percentage of Female Extension Personnel by Geographic Region

Table 64: Frequency of Countries by the Percentage of Female Extension Workers and Total Cereal Yield

Percentage of Female Worker	*Total Cereal Yield (kg/ha)*			*Row Total*	*%*
	< 1,400	1,400-2,500	> 2,500		
< 6	12	11	5	28	34.15
6-12	10	13	5	28	34.15
> 12	6	8	12	26	31.70
Column Total	28	32	22	82	
(%)	34.15	39.02	26.83		100.0

Note—*N* Missing = 32, $\chi^2 = 7.722$, $df = 4$, $p = .1022$, MEF = 6.976.

The results of multiplicative effect are presented in Table 65, indicated that fewer percentage of female extension worker (< 6 per cent) of total extension worker decreases 10.46 per cent total cereal yield over the overall average ($e^{7.4885}$)1782. Having 6-12 per cent of female extension workers of total extension personnel indicated 12.58 per cent decrease in cereal yield and having >12 per cent female extension workers of total extension personnel increases total cereal yield more than 29 per cent.

Table 65: Multiplicative Effect of Percentage of Female Extension (E) Worker (W) on Cereal Yield

Percentage of Female Extension Worker	*Mean*	*SD*	*Effect on Transformed Variables*	*Percentage Increase*	*P*
< 6	7.3750	.6427	–.1105	–1046	
6-12	7.3553	.6035	–.1302	–1258	}.0428
> 12	7.7446	.6283	.2591	2958	
	7.4855	.6423			

Hypothesis 2: The variables, percentage of female extension worker and farmer income are independent of each other. Chi-Square analysis was used to test this hypothesis.

The Chi-Square results are presented in Table 66. The observed relationship between percentage of female extension workers and farmer income was found to be not significant, χ^2 (4, N = 72) = 5.457, p>.24, at the .05 level of significance, therefore, the null hypothesis was accepted. When this analysis was repeated only for the developing countries (per capita GDP of less than US $ 5,000), the observed relationship between these two variables hold the same relationship χ^2 (4, N = 56) = 2.77, p>.59, at the .05 level of significance, therefore, the null hypothesis was accepted.

Table 66: Frequency of Countries by the Percentage of Female Extension Workers and Farmer Income

Percentage of Female Worker	*Farmer Income (US $)*			*Row Total*	*%*
	< 225	225-1,500	> 1,500		
< 6	13	5	6	24	33.33
6-12	11	8	5	24	33.33
> 12	6	8	10	24	33.33
Column Total	30	21	21	72	
(%)	41.6	29.2	29.2		100.00

Note— *N* Missing = 42, χ^2 = 5.457, df = 4, p = .2435, MEF = 7.00. The percentage of row total is rounded.

The results of multiplicative effect are presented in Table 67, indicate that fewer percentage of female extension worker (< 6 per cent) of total extension worker decreases farmer income 31.11 per cent over the overall average ($e^{6.2151}$)500.2. Having 6-12 per cent of female extension workers of total extension personnel indicated 48.07 per cent decrease in farmer income and having >12 per cent female extension workers of total extension personnel increases farmer income more than 159 per cent.

Table 67: Multiplicative Effect of Percentage of Female Extension Worker on Farmer Income

Percentage of Female Extension Worker	*Mean*	*SD*	*Effect on Transformed Variables*	*Percentage Increase*	*p*
< 6	5.8424	2.9873	–.3727	–31.11	
6-12	5.5599	2.4418	–.6552	–48.07	}.0741
> 12	7.1691	2.1460	.9540	159.61	
	6.2151	2.5985			

Hypothesis 3: The variables, percentage of female extension workers and agricultural income/ha of arable land, are independent of each other. Chi-Square analysis was used to test this hypothesis.

The Chi-Square results are presented in Table 68. The observed relationship between percentage of female extension workers and agricultural income/ha of arable land was found to be not significant, χ^2 (4, N = 72) = 5.15, p > .27, at the .05 level of significance, therefore, the null hypothesis was accepted. When this analysis was repeated only for developing countries (per capita GDP of less than US $ 5,000), the observed relationship between these two variables holds the same relationship, χ^2(4, N = 56) = 2.5.65, p. >.61, at the .05 level of significance, therefore, the null hypothesis was accepted.

Table 68: Frequency of Countries by the Percentage of Female Extension Workers and Agricultural Income

Percentage of Female Worker	*Agricultural Income /ha (US $)*			*Row Total*	*%*
	< 525	525-1,500	> 1,500		
< 6	15	5	4	24	33.33
6-12	16	5	3	24	33.33
> 12	9	10	5	24	33.33
Column Total	40	20	12	72	
(%)	55.5	27.9	16.6		100.00

Note—*N* Missing = 42, χ^2 = 5.150, df = 4, p = .2722, MEF = 4.00. The percentage of row total is rounded.

The results of multiplicative effect are presented in Table 69, indicate that fewer percentage of female extension worker (< 6 per cent) of total extension worker decreases agriculture income 17 per cent over the overall average $(e^{5.7268})$306.98. Having 6-12 per cent of female extension workers of total extension personnel indicated 46.38 per cent decrease in agricultural income and having >12 per cent female extension workers of total extension personnel increases agricultural income more than 113 per cent.

Table 69: Multiplicative Effect of Percentage of Female Extension Worker on Agricultural Income

Percentage of Female Extension Worker	*Mean*	*SD*	*Effect on Transformed Variables*	*Percentage Increase*	*p*
< 6	5.5405	2.9924	–.1863	–17.00	
6-12	5.035	2.7361	–.6233	–46.38	}.153
> 12	6.4871	1.5221	.7603	113.89	
	5.7268	2.5055			

Hypothesis 4: The variables, percentage of female extension workers and total fertilizer consumption (kg/ha) are independent of each other. Chi-Square analysis was used to test this hypothesis.

The Chi-Square results are presented in Table 70. The observed relationship between percentage of female extension workers and total fertilizer consumption (kg/ha) was found to be not significant, $\chi^2 (4, N = 80) = 7.44, p = .11$, at the .05 level of significance, therefore the null hypothesis was accepted. When this analysis was repeated only for the developing countries (per capita GDP of less than US $ 5,000), the observed relationship between these two variables hold the same relationship, $\chi^2 (4, N = 67) = 5.16, p > .27$, at the .05 level of significance, therefore, the null hypothesis was accepted.

The results of multiplicative effect are presented in Table 71, indicate that fewer percentage of female extension worker (< 6 per cent) of total extension worker decreases total fertilizer consumption kg/ha arable and permanent crop land 47 per cent over the overall average $(e^{1.9408})$6.96. Having 6-12 per cent of female extension workers of total extension personnel indicated decrease of 46.38 per cent in total fertilizer consumption and having >12 per cent female extension workers of total extension personnel increases total fertilizer consumption more than 113 per cent.

Table 70: Frequency of Countries by the Percentage of Female Extension Workers and Total Fertilizer Consumption

Percentage of Female Worker	*Total Fertilizer Consumption (kg/ha)*			*Row Total*	*%*
	< 25	25-100	> 100		
< 6	15	8	5	28	35.0
6-12	12	8	7	27	33.8
> 12	5	9	11	25	31.2
Column Total	32	25	23	80	
(%)	40.0	31.2	28.8		100.00

Note— *N* Missing = 34, $\chi^2 = 7.4428$, $df = 4$, $p = .1142$, MEF = 7.188.

Table 71: Multiplicative Effect of Percentage of Female Extension Worker on Total Fertilizer Consumption

Percentage of Female Extension Worker	*Mean*	*SD*	*Effect on Transformed Variables*	*Percentage Increase*	*P*
< 6	1.2911	2.1439	–.6497	–47.78	
6-12	1.8113	1.8250	–.1295	–46.38	}.0129
> 12	2.7749	1.3661	.8341	113.89	
	1.9408	1.8964			

Objective 2: *To estimate and compare the quality (educational level) of existing human resources in agricultural extension systems, categorised by geographic region and economic groups and to assess the relationship of this factor with selected agricultural development indicators.*

Level of Education of Extension Personnel

Educational qualification is a primary measure used to assess the quality of extension personnel. With the advancement of science and technology in most developing countries, agricultural extension personnel need increasing levels of education in basic agricultural knowledge, and the application of this knowledge and technology to the solution of practical farm problem, as well as the extension and communication skills to translate this information and skills to farmers.

The data given in Appendix D, TableD-5, indicates that majority of extension personnel in developing countries have very limited education. The percentage of extension personnel with university or higher educational level is presented in Table 72 is comparatively very low in Africa 11.8 per cent, Asia 23.2 per cent and Latin America 31.3 per cent. The percentage of extension personnel with university degree or post-graduate education in Europe was 40.6 per cent and 90.3 per cent in North America. The worldwide average was, however, only 27.9 per cent. Extension personnel at the secondary level are more heterogeneous, with some having completed technical (agricultural) secondary school and other having only general secondary diplomas. In more than a few countries, extension field assistants actually have less than a secondary school diploma. In such countries, agricultural research findings is likely to be beyond the capacity of these inadequately trained extension personnel. Most extension personnel at the secondary level lack sufficient technical training in the agricultural and related sciences to effectively carry out their technology transfer responsibilities.

Table 72: Percentage of Extension Personnel with University Degree by Geographic Region and Economic Group (Mean)

Geographic Region	*Economic Group (PCI)*				
	Low Income	*Lower Middle Income*	*Upper Middle Income*	*High Income*	*Region Mean*
Africa	9.6 (18)	14.9 (9)	23.1(1)	—	11.8 (28)
Asia	13.3 (10)	20.0 (10)	29.2(1)	64.9 (3)	23.2 (24)
Europe	—	—	10.0(1)	46.7 (5)	40.6 (6)
Latin America	12.5 (2)	31.0 (16)	40.0(5)	—	31.3 (23)
North America	—	—	—	90.3 (3)	90.3 (3)
The Near East	33.6 (3)	54.5 (6)	—	34.1 (7)	41.6 (16)
Group Mean	13.1 (33)	28.2 (41)	32.8(8)	52.1(18)	27.9(100)

Note— Number of countries is given in the parentheses, $N = 100$, Missing = 14.

The data in Table 72 reveals that vast majority of extension personnel in Africa, Asia and Latin America are poorly educated (Figure 13) for their extension responsibilities. This situation may be slowing, if not impeding, the transfer of improved technology to farmers. It is planned therefore, to determine empirically if the educational level of extension personnel has any relationship with agricultural development variables. If the relationship between educational level and the agricultural development variables is found to be significant, this finding would lend support to the proposition that governments should improve the quality of their extension personnel through a sustainable in-service education and training programme. In order to test this proposition four null hypotheses were tested using Chi-Square analysis at the .05 level of significance. The percentage of extension personnel with a university or post-graduate degree, or its equivalent will be used as independent variable and four agricultural development indicators as dependent variables.

Hypothesis 1: The variables, percentage of extension personnel with university degree and total cereal yield, are independent of each other.

The Chi-Square results are presented in Table 73. The observed relationship between the percentage of extension personnel with university degree or higher level and total cereal yield was found to be significant, $\chi^2(4, N = 95) = 10.87, p < .05$ at the .05 level of significance, therefore, the null hypothesis was rejected. It appears that an increased percentage of university level educated extension personnel was directly related to increased cereal yields. When this analysis was repeated only for the developing countries (per capita GDP of less than US $ 5,000), the observed relationship between these two variables continued to hold, $\chi^2(4, N = 79) = 11.07, p < .05$, at the .05 level of significance, therefore, the null hypothesis was rejected.

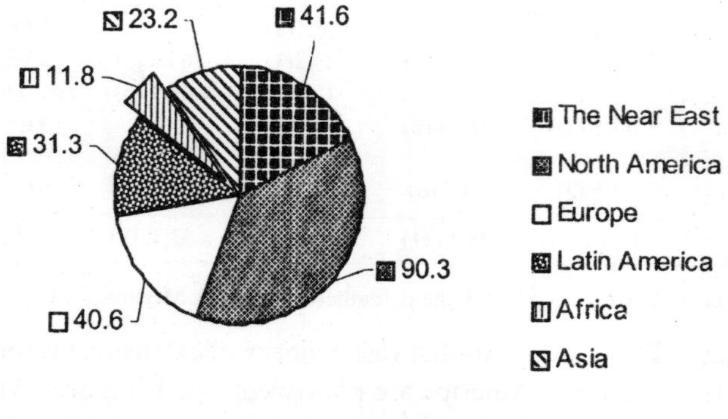

Fig. 13: Percentage of Extension Personnel with University Degree by Region

The results of multiplicative effect analysis are presented in Table 74, indicate that fewer percentage (< 10 per cent) of extension personnel with university degree or higher education decreases total cereal yield 19.11 per cent over the overall average $(e^{7.4679})$1750.9. Having 10-25 per cent of extension workers with university degree or higher degree indicated 16.07 per cent decrease and having >25 per cent extension workers with university degree or higher education indicated increase of more than 35 per cent in total cereal yield.

Table 73: Frequency of Countries by the Percentage of Extension Personnel with University Degree and Total Cereal Yield

Percentage of Extension Personnel with University Degree	*Total Cereal Yield (kg/ha)*			*Row Total*	*%*
	< 1400	1400-2500	> 2500		
< 10	15	12	4	31	32.6
10-25	11	9	7	27	28.4
> 25	6	16	15	37	39.0
Column Total	32	37	26	95	
(%)	33.7	38.9	27.4		100.0

Note— *N* Missing = 14, $\chi^2 = 10.870$, $df = 4$, $p = .02806$, MEF = 7.289.

Table 74: Multiplicative Effect of Percentage of Extension Workers with University Degree on Cereal Yield

Percentage of Extension Personnel with Uni. Degree	*Mean*	*SD*	*Effect on Transformed Variables*	*Percentage Increase*	*p*
< 10	7.2558	.5387	–.2121	–19.11	
10-25	7.2927	.6426	–.1752	–16.07	}.0005
> 25	7.7735	.5819	.3056	35.74	
	7.4679	.6300			

Hypothesis 2: The variables, percentage of extension personnel with university degree or higher education and farmer income, are independent of each other.

The Chi-Square test results are presented in Table 75. The observed relationship between the percentage of extension personnel with university degree or higher education and farmer income was found to be very significant $\chi^2(4, N = 79) = 13.47$, $p<.01$ therefore, the null hypothesis was rejected. It can be inferred from the table that a higher percentage of extension personnel with university degree was related to increased levels of farmer income.

When this analysis was repeated, using only for the developing countries (per capita GDP of less than US \$ 5,000) in the data set, the observed relationship between these two variables was no longer significant, $\chi^2(4, N = 60) = 5.19$, $p > .26$, at the .05 level of significance, therefore, the null hypothesis was accepted.

The result of multiplicative effect presented in Table 76, indicate that fewer percentage (< 10 per cent) of extension personnel with university degree or higher education decreases 55.97 per cent farmer income over the overall average ($e^{6.2806}$) 534.12. Having 10-25 per cent of extension workers with university degree or higher education indicated increase of 12.41 per cent farmer income and having > 25 per cent extension workers with university degree or higher education indicated increase of more than 76 per cent in farmer income.

Table 75: Frequency of Countries by Percentage of Extension Personnel with University Degree and Farmer Income

Percentage of Extension Personnel with University Degree	*Farmer Income (in US $)*			*Row Total*	*%*
	< 225	225-1500	> 1500		
< 10	16	7	3	26	32.9
10-25	4	10	6	20	25.3
> 25	9	10	14	33	41.8
Column Total	29	27	23	79	
(%)	36.7	34.2	29.1		100.0

Note— *N* Missing = 35, $\chi^2 = 13.4705$, $df = 4$, $p = .00919$, MEF = 5.823

Table 76: Multiplicative Effect of Percentage of Extension Personnel with University Degree on Farmer Income

Percentage of Extension Personnel with Uni. Degree	*Mean*	*SD*	*Effect on Transformed Variables*	*Percentage Increase*	*p*
< 10	5.4602	1.4139	−.8204	−55.97	
10-25	6.3976	1.9171	.1170	12.41	}.1194
> 25	6.8479	3.2293	.5673	76.35	
	6.2806	2.4936			

Hypothesis 3: The variables, percentage of extension personnel with university degree and agricultural income/ha of arable land, are independent of each other.

The Chi-Square results are presented in Table 77. The observed relationship between percentage of extension personnel with university degree or higher education and agricultural income was not found to be significant χ^2 (4, N = 79) = 3.96, p >.41 at the .05 level of significance, therefore, the null hypothesis could not be rejected. The minimum expected frequency of two cells out of nine cells was < 5 (22.2 per cent), but still < 25 per cent, therefore, the p value may be accepted. When this analysis was repeated only for developing countries (per capita GDP of less than US $ 5,000), the observed relationship between these two variables continued to be not significant, $\chi^2(4, N = 60) = 3.71$ $p > .44$ at the .05 level of significance, therefore, the null hypothesis was accepted.

Table 77: Frequency of Countries by Percentage of Extension Personnel with University Degree and Agricultural Income

Percentage of Extension Personnel with University Degree	*Agricultural Income (US $)*			*Row Total*	*%*
	< 525	525-1500	> 1500		
< 10	15	9	2	26	32.9
10-25	11	4	5	20	25.3
> 25	18	7	8	33	41.8
Column Total	44	20	15	79	
(%)	55.7	25.3	19.0		100.0

Note—N Missing = 35, $\chi^2 = 3.966$, $df = 4$, $p = .4105$, MEF = 3.79

The results of multiplicative effect presented in Table 78, indicated that fewer percentage (< 10 per cent) of extension personnel with university degree or higher education, increases agricultural income 0.44 per cent over the overall average ($e^{5.8075}$)233.79. Having 10-25 per cent of extension personnel with university degree or higher education indicated increase of 50.32 per cent in agricultural income and having > 25 per cent extension personnel with university degree indicated decrease of more than 21 per cent in agricultural income.

Table 78: Multiplicative Effect of Percentage of Extension Personnel with University Degree on Agricultural Income

Percentage of Extension Personnel with Uni. Degree	*Mean*	*SD*	*Effect on Transformed Variables*	*Percentage Increase*	*p*
< 10	5.8119	1.3645	.0044	.44	
10-25	6.2151	1.8254	.4076	50.32	}.6711
> 25	5.5674	3.2524	-.2401	-21.35	
	5.8075	2.4198			

Hypothesis 4: The variables, percentage of extension personnel with university degree or higher education and total fertilizer consumption (kg/ha) arable and permanent cropland, are independent of each other. Chi-Square analysis was used to test this hypothesis.

The Chi-Square results of this analysis are presented in Table 79. The observed relationship between the percentage of extension personnel with university degree or higher education and total fertilizer consumption in kg/ha arable and permanent cropland was found to be significant, $\chi^2(4, N = 92) = 12.0, p < .05$ at the .05 level of significance, therefore, the null hypothesis was rejected. When this analysis was repeated only for developing countries (per capita GDP of less than US $ 5,000), the observed relationship between these two variables was found to be not significant, $\chi^2(4, N = 75) = 2.14, p > .70$, at the .05 level of significance, therefore, the null hypothesis was accepted.

Table 79: Frequency of Countries by Percentage of Extension Personnel with University Degree and Total fertilizer Consumption

Percentage of Extension Personnel with University Degree	*Fertilizer Consumption (kg/ha)*			*Row Total*	*%*
	< 25	25-100	> 100		
< 10	16	7	5	28	30.4
10-25	11	9	5	25	27.2
> 25	7	17	15	39	42.4
Column Total	34	33	25	92	
(%)	37.0	35.9	27.1		100.0

Note— *N* Missing = 22, χ^2 = 12.0, $df = 4$, p = .0173, MEF = 6.793.

The result of multiplicative effect analysis presented in Table 80, indicated that fewer percentage (< 10 per cent) of extension personnel with university degree or higher education decreases 52.26 per cent total fertilizer consumption (kg/ha arable and permanent cropland) over the overall average ($e^{1.9399}$) 6.96. Having 10-25 per cent of extension workers with university degree or higher education, indicated decrease of 31.09 per cent in total fertilizer consumption and having >25 per cent extension personnel with university degree or higher education indicated increase of more than 111 per cent in total fertilizer consumption.

Objective 3: *To determine and compare the amount of financial resources being invested in agricultural research and extension systems,*

categorised by geographic region and economic groups and to determine their relationship with selected agricultural development indicators.

Table 80: **Multiplicative Effect of Percentage of Extension Personnel with University Degree on Fertilizer Consumption**

Percentage of Extension Personnel with Uni. Degree	*Mean*	*SD*	*Effect on Transformed Variables*	*Percentage Increase*	*p*
< 10	1.2005	1.6311	–.7394	–52.26	
10-25	1.5676	1.8182	–.3723	–31.09	}.0006
> 25	2.6903	1.1467	.7504	111.78	
	1.9399	1.7261			

Allocation of financial resources is another factor that affects the capacity of a country to organise effective agricultural research and extension programmes. It is assumed or hypothesised that the level of investment in agricultural research and extension is also related to the agricultural development process. Therefore, the relationship between agricultural research and extension expenditure and selected agricultural development indicator of responding countries is examined under this objective.

Research Expenditure

The generation of appropriate agricultural technology is considered to be directly related to the agricultural development process. It has been mentioned by a number of scholars that due to the lack of adequate financial resources for agricultural research, the recruitment and retention of competent scientists, the purchase of appropriate scientific equipment, the construction of suitable laboratories, and the advanced training of scientists is frequently an important constraint that limits system effectiveness, therefore, it is important to know what the current level of financial investment is to agricultural research in different countries and then to test the relationship of these variables with selected agricultural development indicators. Therefore, first, agricultural research expenditure data are summarised and presented by geographic region and economic groups in Table 81. Then the relationship of this variable with selected agricultural development variables is tested using four null hypotheses.

The purpose of Table 81 is to compare research expenditure across the geographic region and economic groups. It is evident from the table that among the geographic regions, North America expends the highest amount (U.S. $ 1,760.9 million), Africa (U.S. $ 204.68 million), and the Near East (U.S. $ 118.21 million) Figure 14. Among the economic groups, the highest investment in agricultural research was in high income group (U.S. $ 4,353.97 million), and upper middle income (U.S. $ 279.77 million) group. The worldwide expenditure was U.S. $ 5,430.69 million.

Table 81: Average Research Expenditure by Geographic Region and Economic Group (million)

Geographic Region	*Economic Group (PCI)*				
	Low Income	*Lower Middle Income*	*Upper Middle Income*	*High Income*	*Region Total*
Africa	95.31(18)	109.37(9)	—	—	204.68(27
Asia	280.10(10)	99.01(8)	30.8(1)	1302.83(3)	1712.74(19
Europe	—	—	14.63(1)	1248.36(10)	1262.99(11
Latin America	0.61 (1)	136.22(15)	234.34(5)	—	371.17(21
North America	—	—	—	1760.90(2)	1760.90 (2
The Near East	1.35 (2)	74.98 (5)	—	41.88(4)	118.21(11
Group Mean	377.37(33)	419.58(37)	279.77(7)	4353.97(19)	5430.69(91

Note — Number of countries is given in the parentheses, N=91, Missing = 23. Data used in this analysis were taken from the ISNAR Agricultural Research Indicator Series using 1985 data (actual or extrapolated).

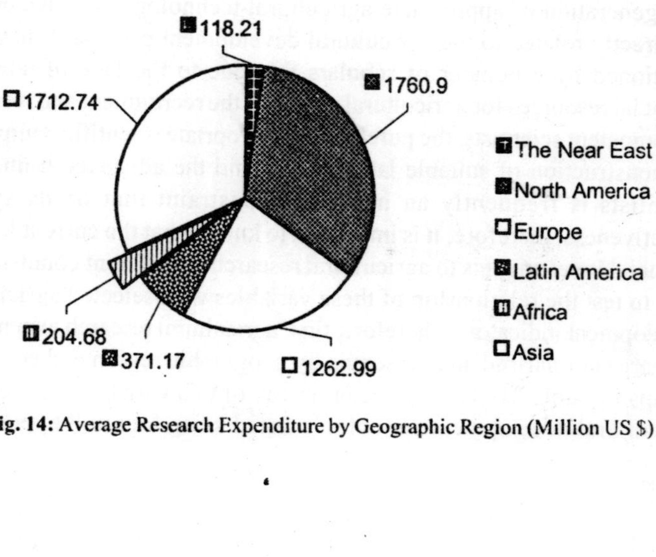

Fig. 14: Average Research Expenditure by Geographic Region (Million US $)

Table 82 presents the research expenditure per farmer. The purpose of Table 82 is to compare the research expenditure per farmer across the geographic regions and economic groups. Among the geographic regions, the maximum research expenditure per farmer was found in North America ($ 498.7) followed by Europe ($ 169.37). In comparison to North America and Europe, the research expenditure per farmer in Africa ($ 2.04), Asia $2.10), the Near East ($ 4.54), and Latin America ($ 9.61) was found to be very low. Among the economic groups, the maximum research expenditure per farmer was high in high income group ($ 265.57) followed by upper middle income ($ 12.21) and lower middle income ($ 4.83). The lowest research expenditure per farmer was in the low income ($ 0.44) group. The worldwide average of research expenditure per farmer was U.S. $ 5.48.

Table 82: Research Expenditure per Farmer by Geographic Region and Economic Group (US $)

Geographic Region	*Economic Group (PCI)*				
	Low Income	*Lower Middle Income*	*Upper Middle Income*	*High Income*	*Region Total*
Africa	1.12	7.73	—	—	2.04
Asia	0.36	3.12	6.34	261.61	2.10
Europe	—	—	10.65	208.62	169.37
Latin America	0.32	6.33	15.42	-	9.61
North America	—	—	—	498.70	498.70
The Near East	0.29	3.87	—	22.04	4.54
Group	0.44	4.83	12.21	265.57	5.48

Note— The research expenditure per farmer presented in this table was computed by dividing the average research expenditure presented in Table 81 by the number of farmers presented in Appendix D (Table D-3), regardless of the number of countries. Therefore, the row and column total of this table do not add up to the total.

Since it has already observed by numerous authors that agricultural researcher cannot perform properly due to lack of financial resources, it is, therefore, important to examine statistically whether or not research expenditure per farmer is related with agricultural development. In order to test this proposition, four null hypotheses were tested using Chi-Square. The agricultural expenditure per farmer is used as independent variable and four agricultural development variables as dependent variables. The following four null hypotheses were tested to assess the potential relationship between agricultural expenditure per farmer and four agricultural development indicators; total cereal yield, farmer income, agricultural income and total fertilizer consumption, at the .05 level of significance.

Hypothesis 1: The variables, research expenditure per farmer and total cereal yield are independent of each other. Chi-Square analysis was used to test this hypothesis.

The Chi-Square results are presented in Table 83. The observed relationship between research expenditure per farmer with total cereal yield was found to be very significant, χ^2 (4, N = 84) = 38.10, p<.0001 at the .05 level of significance, therefore, the null hypothesis that research expenditure per farmer and total cereal yield are independent was rejected. Since the value of one cell count is 0, the χ^2 value may not be trusted but the minimum expected frequency was > 5 (*i.e.* 5.71), therefore the p value may be trusted. When this analysis was repeated only for the developing countries (per capita of less than US $ 5,000), the observed relationship between these two variables does not hold the same relationship, χ^2(4, N = 66) = 6.09, p>.05, at the .05 level of significance. Therefore, the null hypothesis was accepted.

Table 83: Frequency of Countries by Research Expenditure per Farmer and Total Cereal Yield

Research Expenditure per Farmer in U.S.$	*Total Cereal Yield (kg./ha)*			*Row Total*	*%*
	< 1400	1400-2500	> 2500		
< 5	20	15	4	39	46.4
5-20	7	11	2	20	23.8
> 20	0	7	18	25	29.8
Column Total	27	33	24	84	
(%)	32.1	39.3	28.6		100.0

Note — N Missing = 30, χ^2 = 38.1013, df = 4, p = .00000, MEF = 5.714.

The results of multiplicative effect analysis are presented in Table 84, indicate that fewer research expenditure per farmer (< 5 US$) increases total cereal yield 12.15 per cent over the overall average ($e^{7.2514}$)1410.1. Having 5-20 US $ research expenditure per farmer indicated 26.29 per cent decrease and having >20 US $ per cent research expenditure per farmer indicated increase of more than 57 per cent in total cereal yield.

Table 84: Multiplicative Effect of Research Expenditure per Farmer on Cereal Yield

Research Expenditure per Farmer	*Mean*	*SD*	*Effect on Transformed Variables*	*Percentage Increase*	*p*
< 5	7.3661	.4249	.1147	12.15	
5-20	6.9464	.3546	–.3050	–26.29	}.0533
> 20	7.7075	.0426	.4561	57.79	
	7.2514	.4452			

Hypothesis 2: The variables, research expenditure per farmer and farmer income, are independent of each other.

Chi-Square test results are presented in Table 85. The observed relationship between the research expenditure per farmer and farmer income was found to be very significant χ^2 (4, N = 76) = 55.79, p<.0001, at the .05 level of significance. Therefore, the null hypothesis that research expenditure per farmer with farmer income are independent was rejected. It can be inferred from the table that as the research expenditure per farmer increases the farmer income also increases. Since the value of one cell is 0, the χ^2 value cannot be trusted. The minimum expected cell frequency of only one cell is 5 (4.921), that is 11.1 per cent (< 25 per cent), therefore, the p value may be trusted. When this analysis was repeated, using only for the developing countries (per capita GDP of less than US $ 5,000), the observed relationship between these two variables holds the same relationship, χ^2 (4, N = 54) = 19.99, p<.001, at the .05 level of significance, therefore, the null hypothesis was rejected.

Table 85: Frequency of Countries by Research Expenditure per Farmer and Farmer Income

Research Expenditure per Farmer	*Farmer Income (in US $)* < 225	225-1500	> 1500	*Row Total*	*%*
< 5	24	12	0	36	47.4
5-20	4	10	3	17	22.4
> 20	1	3	19	23	30.2
Column Total	29	25	22	76	
(%)	38.2	32.9	28.9		100.0

Note — *N* Missing=38, χ^2 =55.7936, df = 4, p = .00000, MEF = 4.921, cells with expected frequency <5 -1 of 9 (11.1%).

The result of multiplicative effect presented in Table 86, indicate that fewer research expenditure per farmer (< 5 US$) increases farmer income 37.84 per cent over the overall average ($e^{5.2439}$) 189.41. Having 5-20 US $ research expenditure per farmer indicated decrease of 19.01 per cent farmer income and having > 20 US $ research expenditure per farmer indicated decrease of more than 15 per cent in farmer income.

Table 86: Multiplicative Effect of Research Expenditure per Farmer on Farmer Income

Research Expenditure per Farmer	*Mean*	*SD*	*Effect on Transformed Variables*	*Percentage Increase*	*p*
< 5	5.5648	.2853	.3209	37.84	
5-20	5.0331	.7601	–.2108	–19.01	}.3139
> 20	5.0741	.0561	–.1658	–15.62	
	5.2439	.58'4			

Hypothesis 3: The variables, research expenditure per farmer and agricultural income/ha of arable land, are independent of each other.

The Chi-Square results are presented in Table 87. The observed relationship between research expenditure per farmer with agricultural income was found to be very significant χ^2 (4, N = 76) = 30.62, p<.0001 at the .05 level of significance. Therefore, the null hypothesis that research expenditure per farmer and agricultural income/ha of arable land are independent was rejected. Since the minimum expected frequency was < 5 (*i.e.* 3.132) for three cells (*i.e.* 33.3 per cent), therefore, the p value may not be trusted. When this analysis was repeated only for developing countries (per capita GDP of less than US $ 5,000), the observed relationship between these two variables does not hold the same relationship, χ^2(4, N = 54) = 5.25, p >.26, at the .05 level of significance, therefore, the null hypothesis was accepted.

The results of multiplicative effect presented in Table 88, indicated that fewer research expenditure per farmer (< 5 US $), increases agricultural income 74.46 per cent over the overall average ($e^{5.5421}$)255.21. Having 5-20 U.S. $ research expenditure per farmer indicated decrease of 11.25 per cent in agricultural income, and having > US $ 20 research expenditure per farmer decreases agricultural income more than 62 per cent.

Table 87: Frequency of Countries by Research Expenditure per Farmer and Agricultural Income

Research Expenditure per Farmer in U.S. $	*Agricultural Income (US $)*			*Row Total*	%
	< 525	525-1500	> 1500		
< 5	28	7	1	36	47.4
5-20	12	4	1	17	22.4
> 20	4	7	12	23	30.3
Column Total	44	18	14	76	
(%)	57.9	23.7	18.4		100.0

Note—*N* Missing = 38, $\chi^2 = 30.6224$, $df = 4$, $p = .00000$, MEF = 3.132, cells with expected frequency < 5-3 of 9 (33.3%).

Table 88: Multiplicative Effect of Research Expenditure per Farmer on Agricultural Income

Research Expenditure per Farmer in US$	*Mean*	*SD*	*Effect on Transformed Variables*	*Percentage Increase*	*p*
< 5	6.0986	.4905	.5565	74.46	
5-20	5.4227	1.6331	–.1194	–11.25	}.4018
> 20	4.9094	2.1293	–1.0327	–62.40	
	5.5421	1.3723			

Hypothesis 4: The variables, research expenditure per farmer and total fertilizer consumption (kg/ha) arable and permanent cropland, are independent of each other. Chi-Square analysis was used to test this hypothesis.

The Chi-Square results of this analysis are presented in Table 89. The observed relationship between the extension expenditure per farmer and total fertilizer consumption in kg/ha arable and permanent cropland was found to be very significant, χ^2 (4, $N = 84$) = 35.27, $p<.0001$ at the .05 level of significance. Therefore, the null hypothesis that research expenditure per farmer and total fertilizer consumption are independent was rejected. When this analysis was repeated only for developing countries (per capita GDP of less than US $ 5,000), the observed relationship between these two variables holds the same relationship, χ^2 (4, $N = 65$) = 11.64, $p < .05$, at the .05 level of significance, therefore, the null hypothesis was rejected.

Table 89: **Frequency of Countries by Research Expenditure per Farmer and Total Fertilizer Consumption**

Research Expenditure per Farmer (US $)	*Fertilizer Consumption (kg/ha)*			*Row Total*	*%*
	<25	25-100	>100		
<5	21	11	5	37	44.0
5-20	7	11	3	21	25.0
>20	1	6	19	26	31.0
Column Total	29	28	27	84	
(%)	34.5	33.3	32.2		100.0

Note— *N* Missing = 30, $\chi^2 = 35.2761$, $df = 4$, p = .00000, MEF = 6.75

The result of multiplicative effect analysis are presented in Table 90, indicated that fewer research expenditure per farmer (< U.S. $5) increases total fertilizer consumption 10.27 per cent over the overall average($e^{.8574}$) 2.36. Having U.S. $ 5-20 research expenditure per farmer, indicated decrease of 49.78 per cent in total fertilizer consumption and having > U.S. $ 20 research expenditure per farmer indicated increase of more than 433 per cent in total fertilizer consumption.

Table 90: Multiplicative Effect of Research Expenditure per Farmer on Fertilizer Consumption

Research Expenditure per Farmer US$	*Mean*	*SD*	*Effect on Transformed Variables*	*Percentage Increase*	*P*
<5	.9592	1.7987	.0978	10.27	
5-20	.1686	1.8062	–.6888	–49.78	}.2808
>20	2.5326	.7703	1.6752	433.99	
	.8574	1.7905			

Extension Expenditure

As discussed earlier, adequate extension expenditure is assumed to be important for the effective operation of a national extension system. It has also been reported by numerous extension scholars that due to inadequate financial investment to extension, the recruitment and retention of competent extension personnel, plus adequate provision for in service training, transportation, housing and the conduct of extension programmes cannot be properly carried out. To provide an overview of national investment in agricultural extension system, it is useful to compare level of financial expenditure to extension across the geographic regions and different economic groups. In addition, to test the assumptions made by extension scholars about the extension expenditure, it is necessary to determine if agricultural extension expenditure is related with selected agricultural development indicators. If it is found that extension expenditure is significantly related with agricultural development, then it would provide empirical support for countries to increase their investment in extension. Keeping these factors in mind, cross country comparisons of financial investment in extension is presented in Table 91 and then its relationship with selected agricultural development indicators is tested at .05 level of significance using four null hypotheses.

Table 91: Total Extension Expenditure by Geographic Region and Economic Group (million U.S. $)

Geographic Region	*Economic Group (PCI)*				
	Low Income	*Lower Middle Income*	*Upper Middle Income*	*High Income*	*Region Total*
Africa	75.55(17)	100.73 (7)	11.00(1)	—	187.72(25)
Asia	525.69 (9)	129.95 (8)	89.10(1)	326.43(3)	1071.17(21)
Europe	—	—	23.12(2)	125.77(5)	148.89 (7)
Latin America	6.18 (2)	127.02(13)	275.39(5)	—	409.22(20)
North America	—	—	—	358.00(1)	358.00 (1)
The Near East	2.55 (3)	62.24 (4)	-	83.07(7)	147.86(14)
Group Mean	611.04(31)	419.94(32)	398.61(9)	893.27(16)	2322.86(88)

Note—Number of countries are given in the parentheses.

The purpose of Table 91 was to compare the extension expenditure across the geographic regions and economic groups. It is evident from the table that among the geographic regions, Asia expends highest amount (U.S. $ 1071.17 million) (Figure 15), followed by Latin America (U.S. $ 409.22 million), North America (U.S. $358.0 million), Africa (U.S. $ 187.72 million), Europe (U.S. $ 148.89 million), and the Near East ($ 147.86 million). Among the economic groups, the highest amount was expend by high income group (U.S. $ 893.27 million), followed by low income (U.S. $ 611.04 million), lower middle income (U.S. $ 419.94 million), and upper middle income group (U.S. $ 389.61 million). The total level of expenditure for the 88 countries included in this table was U.S. $ 2,322.86 million. Since the number of extension personnel is also comparatively much higher in Asia, and also in Africa the above expenditure gives information that could be somewhat misleading. Therefore, it is useful to compute a measure that can be used to compare the relative value of extension investment in each country. Extension expenditure per farmer was selected as this measure and these data are presented in Table 92.

Among the geographic regions, North America expends highest amount (U.S. $ 101.39) (Figure 16) per farmer in contrast to Asia (U.S. $ 1.32). The expenditure per farmer in Africa (U.S. $ 1.87), the Near East (U. S. $ 5.68), and Latin America (U. S. $ 10.59) is also relatively low in comparison to North America (U. S. $ 101.39). The worldwide expenditure per farmer was (U.S. $ 2.34).

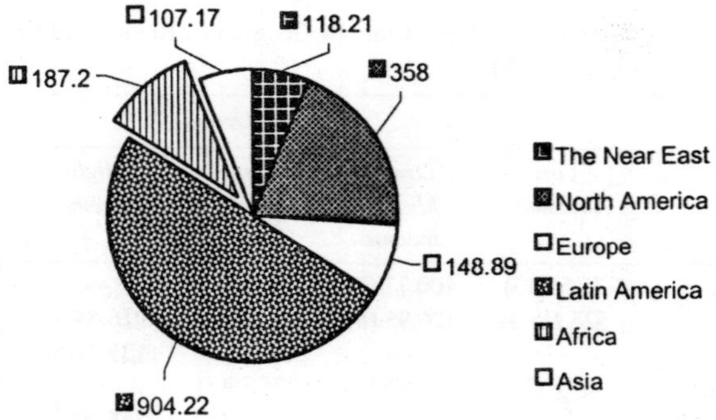

Fig. 15: Extension Expenditure by Geographic Region (Million US $)

Table 92: Average Extension Expenditure per Farmer by Geographic Region and Economic Group (US$)

Geographic Region	*Economic Group (PCI)*				
	Low Income	*Lower Middle Income*	*Upper Middle Income*	*High Income*	*Region*
Africa	0.89	7.12	7.98	—	1.87
Asia	0.68	4.10	4.33	65.55	1.32
Europe	—	—	15.70	21.02	819.97
Latin America	3.56	5.90	18.12	—	10.59
North America	—	—	—	101.39	101.39
The Near East	0.54	3.21	—	43.72	5.68
Group	0.71	4.84	17.40	54.48	2.34

Note—The extension expenditure per farmer presented in this table was computed directly by dividing the total extension expenditure given in Table 91, by the total number of farmers given in Appendix D (Table D-3), regardless of the number of countries. Therefore, the row and column totals in this table do not add to the total.

In order to assess the relationship of extension expenditure per farmer with agricultural development indicators, the following four null hypotheses were tested. The independent variable used was extension expenditure per farmer and the dependent variables were: total cereal yield kg/ha, farmer income, agricultural income, and total fertilizer consumption (kg/ha) at the .05 level of significance.

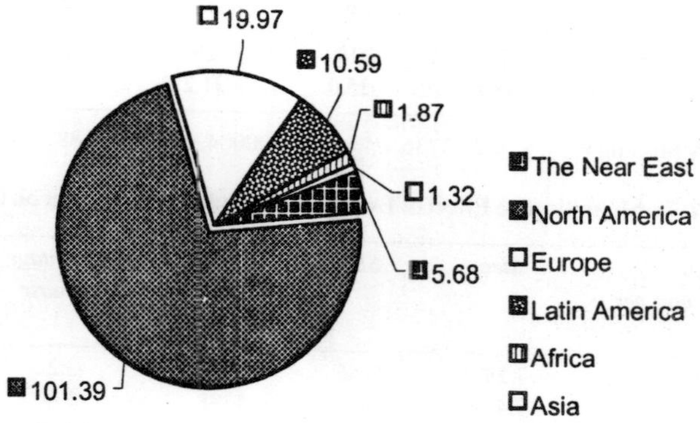

Fig. 16: Average Extension Expenditure per Farmer by Geographic Region (US$)

Hypothesis 1: The variables, extension expenditure per farmer and total cereal yield are independent of each other. Chi-Square analysis was used to test this hypothesis.

The Chi-Square results are presented in Table 93. The observed relationship between extension expenditure per farmer with total cereal yield was found to be very significant, χ^2 (4, N = 80) = 25.37, p<.0001 at the .05 level of significance. Therefore, the null hypothesis that extension expenditure per farmer and total cereal yield are independent was rejected. When this analysis was repeated only for the developing countries (per capita GDP of less than US $ 5,000), the observed relationship between these two variables does not hold the same relationship, χ^2 (4, N = 66) = 4.10, p>.39, at the .05 level of significance. Therefore, the null hypothesis was accepted.

The results of multiplicative effect analysis are presented in Table 94, indicate that fewer extension expenditure per farmer (< 3 US$) decreases total cereal yield 3.22 per cent over the overall average ($e^{7.3198}$)1509.9. Having 3-16 (US$) extension expenditure per farmer indicated 15 per cent increase in cereal yield and having >16 (US$) extension expenditure per farmer indicated decrease of about 5 per cent in total cereal yield.

Table 93: Frequency of Countries by Extension Expenditure per Farmer and Total Cereal Yield

Extension Expenditure per Farmer in US $	*Total Cereal Yield (kg/ha)*			*Row Total*	*%*
	< 1400	1400-2500	> 2500		
< 3	19	8	4	31	38.7
3-16	5	14	7	26	32.5
> 16	3	6	14	23	28.8
Column Total	27	28	25	80	
(%)	33.8	35.0	31.2		100.0

Note— N Missing = 34, $\chi^2 = 25.3736$, $df = 4$, $p = .00004$, MEF = 7.188.

Table 94: Multiplicative Effect of Extension Expenditure per Farmer on Cereal Yield

Research Expenditure per Farmer	*Mean*	*SD*	*Effect on Transformed Variables*	*Percentage Increase*	*p*
< 3	7.2871	.5156	–.0327	–3.22	
3-16	7.4596	.5852	.1398	15.00	}.8670
> 16	7.2686	.9456	–.0512	–4.99	
	7.3198	.6364			

Hypothesis 2: The variables, extension expenditure per farmer and farmer income, are independent of each other. Chi-Square analysis was used to test this hypothesis.

Chi-Square test results are presented in Table 95. The observed relationship between the extension expenditure per farmer and farmer income was found to be very significant $\chi^2 (4, N = 68) = 41.31, p < .0001$, at the .05 level of significance. Therefore, the null hypothesis that extension expenditure per farmer with farmer income are independent was rejected. Although, two cell counts were 0, the minimum expected frequency was more than 5 (*i.e.* 5.029). Hence, the probability can be trusted but not the χ^2 value. When this analysis was repeated, using only for the developing countries (per capita GDP of less than US $ 5,000), the observed relationship between these two variables holds the same relationship, $\chi^2 (4, N = 53) = 31.81, p < .0001$, at the .05 level of significance, therefore, the null hypothesis was rejected.

The result of multiplicative effect are presented in Table 96, indicate that fewer extension expenditure per farmer (< 3 US$) decreases farmer income 42.99 per cent over the overall average ($e^{4.8258}$) 124.69. Having

US $ 3-16 extension expenditure per farmer indicated increase of 51.85 per cent farmer income and having > US $ 16 extension expenditure per farmer indicated increases farmer income more than 238 per cent.

Table 95: Frequency of Countries by Extension Expenditure per Farmer and Farmer Income

Extension Expenditure per Farmer (US $)	*Farmer Income (in US $)*			*Row Total*	%
	< 225	225-1500	> 1500		
< 3	19	5	0	24	35.3
3-16	7	13	5	25	36.8
> 16	0	6	13	19	27.9
Column Total	26	24	18	68	
(%)	38.2	35.3	26.5		100.0

Note—*N* Missing=46, χ^2 =41.3191, df = 4, p = .0000, MEF = 5.029.

Table 96: Multiplicative Effect of Extension Expenditure per Farmer on Farmer Income

xtension xpenditure per armer US$	*Mean*	*SD*	*Effect on Transformed Variables*	*Percentage Increase*	*p*
3	4.2665	3.0743	-.95593	-42.99	
-16	5.2462	.3761	.4204	51.85	}.4787
16	6.0484	.7302	1.2226	238.68	
	4.8258	2.5011			

Hypothesis 3: The variables, Extention expenditure per farmer and agricultural income/ha of arable land, are independent of each other.

The Chi-Square results are presented in Table 97. The observed relationship between extension expenditure per farmer with agricultural income was found to be very significant χ^2 (4, N = 68) = 23.47, p<.0001 at the .05 level of significance. Therefore, the null hypothesis that extension expenditure per farmer and agricultural income/ha of arable land are independent was rejected. Since one cell count has a 0 value, and the minimum expected frequency was less than 5 (*i.e.* 3.35) in four cells (*i.e.* 44.4 per cent), both the χ^2 and p value can not be trusted. When this analysis was repeated only for developing countries (per capita GDP of less than US $ 5,000), the observed relationship between these two variables does not hold the same relationship, χ^2 (4, N = 53) = 4.31,

$p > .36$, at the .05 level of significance, therefore, the null hypothesis was accepted.

The results of multiplicative effect presented in Table 98, indicated that fewer extension expenditure per farmer (< 3 US $), decreases agricultural income 74.32 per cent over the overall average ($e^{4.8656}$)129.75. Having US $ 3-16 extension expenditure per farmer indicated increases agriculture income 110.29 per cent, and having > US $ 16 extension expenditure per farmer increases agricultural income more than 233 per cent.

Table 97: Frequency of Countries by Extension Expenditure per Farmer and Agricultural Income

Extension Expenditure per Farmer in U.S. $	*Agricultural Income (US $)*			*Row Total*	*%*
	< 525	525-1500	< 1500		
< 3	17	7	0	24	35.3
3-16	17	6	2	25	36.8
> 16	5	4	10	19	27.9
Column Total	39	17	12	68	
(%)	57.34	25.0	17.64		100.0

Note— *N* Missing = 46, $\chi^2 = 23.4721$, $df = 4$, $p = .0001$, MEF = 3.353 <5, cells with expected frequency < 5-4 of 9 (44.4%).

Table 98: Multiplicative Effect of Extension Expenditure per Farmer on Agricultural Income

Extension Expenditure per Farmer US$	*Mean*	*SD*	*Effect on Transformed Variables*	*Percentage Increase*	*p*
< 3	4.2246	3.5514	–.6410	-47.32	
3-16	5.6089	.6605	.7433	110.29	}.5219
> 16	6.0709	1.3356	1.2053	233.78	
	4.8646	2.9138			

Hypothesis 4: The variables, extension expenditure per farmer and total fertilizer consumption (kg/ha) arable and permanent cropland, are independent of each other. Chi-Square analysis was used to test this hypothesis.

The Chi-Square results of this analysis are presented in Table 99. The observed relationship between the extension expenditure per farmer and total fertilizer consumption in kg/ha arable and permanent cropland was found to be very significant, χ^2 (4, N = 80) = 29.16, p<.0001 at the .05 level of significance. Therefore, the null hypothesis that extension expenditure per farmer and total fertilizer consumption are independent was rejected. When this analysis was repeated only for developing countries (per capita GDP of less than US $ 5,000), the observed relationship between these two variables holds the same relationship, χ^2 (4, N = 64) = 13.39, p < .01, at the .05 level of significance, therefore, the null hypothesis was rejected.

Table 99: Frequency of Countries by Extension Expenditure per Farmer and Total Fertilizer Consumption

Extension Expenditure per Farmer (US $)	*Fertilizer Consumption (kg/ha)*			*Row Total*	*%*
	< 25	*25-100*	*> 100*		
< 3	21	6	2	29	44.0
3-16	4	13	8	25	25.0
> 16	4	9	13	26	31.0
Column Total	29	28	23	80	
(%)	36.2	35.0	28.8		100.0

Note— N Missing = 34, χ^2 = 34.1681, df = 4, p = .00001, MEF = 7.188.

The result of multiplicative effect analysis are presented in Table 100, indicated that fewer extension expenditure per farmer (< U.S. $3) decreases total fertilizer consumption 40.64 per cent over the overall average ($e^{1.6907}$) 5.42. Having US $3-16 extension expenditure per farmer, indicated increase of 68.98 per cent in total fertilizer consumption and having > US $ 16 extension expenditure per farmer indicated increase of more than 6 per cent in total fertilizer consumption.

Table 100: Multiplicative Effect of Extension Expenditure per Farmer on Fertilizer Consumption

Extension Expenditure per Farmer US$	*Mean*	*SD*	*Effect on Transformed Variables*	*Percentage Increase*	*p*
< 3	1.1692	1.6214	–.5215	–40.64	
3-16	2.2153	1.1159	.5246	68.98	}.1712
> 16	2.4179	.6678	.7272	06.93	
	1.6907	1.4304			

Objective 4: *To assess the relationship between the agricultural development indicators and measures of the human and financial resources being invested in national agricultural research and extension systems.*

Four agricultural development indicators (total cereal yields, farmer income, agricultural income, and total fertilizer consumption) were used in this analysis. The primary human resource indicators were the following: number of farmers (economically active population in agriculture) per extension worker, arable land/ha per extension worker, number of farmers per researcher, arable land (ha) per researcher, and ratio of extension workers per researcher. The other human resource variables were percentage of subject-matter specialists to the total number of extension personnel, the percentage of extension personnel with university or post-graduate level education, and the percentage of female extension personnel to the total number of extension personnel.

Since data on financial resources was incomplete for a large number of countries, the regression analysis for financial resources was conducted separately and follows the human resource analysis. As mentioned at the outset of the chapter, two sets of regression analysis are presented under this objective. The first set includes all 114 responding countries and the second set includes those responding countries that have per capita GDP of less than U.S. $ 5,000. The same variables, however, were used in both sets of analysis. Under this objective, the following null hypotheses were tested to assess the relationship of agricultural development indicator variables with these human and financial resource variables at the .05 level of significance.

Hypothesis 1: There is no relationship between selected human resources variables and total cereal yield. This hypothesis was tested using multiple regression analysis.

The main concern of this analysis was to statistically examine the relationship between the dependent variable, total cereal yield, with the five independent variables in the model. This analysis is also used to estimate the magnitude of the relationship between the dependent variables and best linear combination of independent variables. It has been posited by extension scholars that the fewer number of farmers per extension workers, higher educational level of extension personnel, fewer hectares of arable land per extension worker, and fewer number of farmers per researcher enhances the effectiveness of agricultural research and extension systems and hence, contributions to agricultural development.

Table 101 presents one tailed inter-correlation matrix of the dependent variable, total cereal yield in kg/ha and five independent variables: number of farmers per researcher, arable land per extension worker, percentage of subject-matter specialists to total extension personnel, percentage of extension personnel with university degree or higher education (educational level), and number of farmers per extension worker. These independent variables were selected after multicollineary diagnosis. The variables which were found to be multicollinear were not included in the model.

Table 101: Inter-correlation Matrix Among Selected Human Resource Variables with Total Cereal Yield as a Dependent Variable (Worldwide Data)

	Variable	X_1	X_2	X_3	X_4	X_5
X_1.	No. of Farmers/Researcher	1.000				
X_2.	Arable land/Extension worker	–.049				
X_3.	% of Subject Matter Specialist	–.093	.212			
X_4.	Educational Level	–.261	.275*	.419***		
X_5.	No. of Farmers/ Extension Worker	.533***	.082	.170	–.071	
Y.	Cereal Yield kg/ha	–.281*	–.064	.276**	.400***	–.273

Note— $N = 83$. *$p < .05$, **$p < .01$, ***$p < .001$.

These five variables are free from multicollinearity. Out of these five independent variables four were found correlated at different degrees of significance with the dependent variable, total cereal yield in kg/ha. The variable, percentage of extension personnel with university degree or higher level of education was the most positively correlated ($r = .400$***, $p < .001$) with cereal yield among all human resource variables. The variable, percentage of subject-matter specialist of total extension personnel was also positively correlated ($r = .276$**, $p < .01$) with cereal yield. The other two variables; the number of farmers per extension worker ($r = -.273$**, $p < .01$) and the number of farmers per researcher ($r = -.281$**, $p < .01$) in the model, were negatively correlated as expected. The fifth variable; arable land per extension worker was not found significant and indicates negative direction.

In practical terms, the positive correlation of the independent variables, percentage of extension personnel with university degree or higher education and percentage of subject matter specialists, indicates that if percentage of these variables is increasing then cereal yield also tend to increase. The negative direction of the two significant independent variables in the model indicates that if the number of farmers per extension worker and researcher is decreasing (*i.e.* research and extension intensity is increasing), then cereal yield also tend to increase. When this analysis was repeated only for those countries having per capita GDP of less than US $ 5,000, a similar direction in the relationship was found between the independent and dependent variables. However, only the variabl^ number of farmers per researcher ($r = -.286^{**}$, $p < .001$) was found significantly and negatively correlated with total cereal yield.

Table 102 shows the standardised coefficients of five independent variables with respect to their influence on the total cereal yield. It reveals that variables, percentage of extension personnel with university degree ($\beta = .329$, $p < .01$) and number of farmers per extension worker ($\beta = -41$, $p < .05$) had a significant relationship with total cereal yield. Therefore, this analysis indicates that higher percentage of extension workers with university degree and fewer number of farmers per extension worker are directly related with higher cereal yield in kg/ha. The variable, percentage of subject matter specialists (SMS) ($\beta = .212$, $p > .05$) and arable land per extension worker ($\beta = -.182$, $p > .05$) are approaching significance at the .05 level and may also be related, but this relationship could not be established at minimum confidence level.

Table 102: Multiple Regression Coefficient with Cereal Yield as Dependent Variable

	Variable	*B*	*T*	*Beta*	*Significance*
X_1.	No. of Farmers/Researcher	–.001	–.47	–.056	.6396
X_2.	Arable Land/Ext. Worker	–.004	–1.80	–.182	.0748
X_3.	% of Subject Matter Specialist	.966	1.93	.212	.0570
X_4.	Educational Level	15.052	2.94	.329	.0043
X_5.	No. of Farmers/Ext. Worker	–.073	–2.03	–.241	.0453
	(Constant)	1560.661	7.35		.0000

When this analysis was repeated for those countries having per capita GDP of less than US $ 5,000, the relationship of Beta values of all independent variables with the dependent variables was found to be similar. Although none of the independent variables were found to be significant at the .05 level. Three variables, including the percentage of subject matter specialists of total extension personnel

($\beta = .1937, p =.1463$) percentage of extension personnel with university degree ($\beta = .1928, p = .1524$), and arable land per extension worker ($\beta = -.1981, p = .1822$), were approaching significance at the .10 level.

Table 103 indicates that, using the worldwide data set, 23.8 per cent of the variance in the total cereal yield was explained by the number of farmers per researcher, arable land per extension worker, percentage of subject matter specialists of total extension personnel, percentage of extension personnel with university degree, and number of farmers per extension worker. The level of significance, $p = <.0001$, indicates that the results are acceptable. In other words, there is a 95 per cent probability that 23.8 per cent of the variance in the total cereal yield is due to the five variables in the model. Hence, the null hypothesis, that there is no relationship between these human resource variables and cereal yield was rejected at the .05 level of significance.

Table 103: The Overall F-Test

Multiple *R*	.533	Analysis of variance	*D.F.*	*F*	*p*
R^2	.284	Regression	5	6.12	$p<.05$
Adjusted *R* Square	.238	Residual	77		
Standard Error	1094.689				
Durbin Watson Test	1.321				

When this analysis was repeated for those countries having per capita GDP of less than US \$ 5,000, only 11.1% of the variance in the total cereal yield was explained by independent variables in the model with $F = 2.751$, and $p = .025$, which was $p < .05$. Hence, the null hypothesis was rejected at the .05 level of significance. In other words, total cereal yield in both sets of analysis was partially dependent on the five variables in the model, however, only the worldwide data set established a significant relationship between the independent and dependent variables in the model.

Hypothesis 2: There is no relationship between selected human resources variables of agricultural research and extension systems and farmer income. This hypothesis was tested using multiple regression analysis.

Table 104 presents the inter-correlation matrix of five independent variables with farmer income as the dependent variables. Of the five independent variables, percentage of extension personnel with a university degree or higher education ($r = .478^{***}, p < .001$), percentage of subject matter specialist ($r = .422^{***}$, $p < .001$) and arable land per extension worker (r = –.369***, p <.001) were positively and significantly correlated with farmer income. This result indicates that if the percentage of extension personnel with a university degree or higher education, the percentage of subject matter specialists, and arable land per extension worker is increased, then farmer income also tends to increase. The variables, number of farmers per researcher (r = –.243) and number of farmers per extension workers ((r = –.145) are not significantly correlated with farmer income but indicate a negative direction.

Table 104: Inter-correlation Matrix Among Selected Human Resource Variables with Farmer Income (U.S.$) Dependent Variable (Worldwide Data)

	Variable	X_1	X_2	X_3	X_4	X_5
X_1.	No. of Farmers/Researcher	1.000				
X_2.	Arable land/Extension worker	–.095				
X_3.	% of Subject Matter Specialist	.004	.216			
X_4.	Educational Level	–.346	.284	.411***		
X_5.	No. of Farmers/Extension Worker	.370***	.093	.251	–.016	
Y.	Cereal Yield kg/ha	–.243	.369***	.422***	.478***	–.145

Note— N = 73. **p < .01, ***p < .001.

When this analysis was repeated with the low income countries having per capita GDP of less than U.S. $ 5,000, only the variable, percentage of extension personnel with university degree or higher education (r = .3299**, p < .01), was found to be significantly correlated with farmer income. The variable, percentage of subject matter specialists indicates a positive direction, but was not significant. The other three variables, number of farmers per extension worker, (r = –.1855), number of farmers per researchers (r = –.172) and arable land per extension worker (r = –.1728) in the model, indicate a negative direction and were not significant.

To examine the influence of variables in the model, a multiple regression analysis was conducted with farmer income as the dependent variable. The purpose of this analysis was to determine which extension and research variables were significantly related with farmer income. Table 105 presents the standardised coefficient of five independent variables with respect to their influence of the dependent variable farmer income. It is evident from this table that the variables, percentage of subject matter specialists, percentage of extension personnel with university degree or higher education, arable land per extension worker and number of farmers per extension worker were significant in explaining farmer income. Only the variable, number of farmers per research was not significantly related to farmer income at the .05 level of significance.

Table 105: Multiple Regression Coefficient with Farmer Income (U.S.$)

	Variable	*B*	*T*	*Beta*	*Signfi.*
X_1	No. of Farmers/Researcher	–.012	–.449	–.049	.6551
X_2	Arable Land/Ext. Worker	.034	2.424	.243	.0180
X_3	% of Subject Matter Specialist	155.472	2.933	.321	.0046
X_4	Educational Level	67.609	2.223	.256	.0296
X_5	No. of Farmers/Ext. Worker	–.436	–2.119	–.226	.0378
	(Constant)	24.117	.019		.9852

When this analysis was repeated for those countries having per capita GDP of less than US $ 5,000/, the variables, arable land per extension worker (β=.3747, p>.05) and percentage of extension personnel with university degree or higher education (β = .4465, p<.001) were found significant in explaining the farmer income. The other three variables; number of farmers per extension worker (β = –1.969, p>.05), percentage of subject matter specialists (β = .047, p>.05), and number of farmers per researcher (β = –.1213, p>.05) were not significantly related at the .05 level of significance.

Table 106 indicates that 34.7 per cent of variance in the farmer income is explained by the variables; number of farmers per researcher, arable land per extension worker, percentage of subject matter specialists, percentage of extension personnel with university degree or higher education, and number of farmers per extension worker. In other words, there is at least 95 per cent probability that 34.7 per cent of the variance in the farmer income is due to the five independent variables in the model. Hence, the null hypothesis that there is no relationship between the selected human resources variables and farmer income is rejected at the .05 level of significance.

When this analysis was repeated with the countries having per capita GDP of less than US $ 5,000, 16.7 per cent of the variance in the farmer income was explained by these variables collectively. The level of significance of F-test was p = .014, hence the null hypothesis was rejected.

In other words, farmer income in both sets of analysis is partially dependent on the five variables in the model collectively.

Table 106: The Overall F-Test

Multiple *R*	.627	Analysis of variance	*D.F.*	*F*	*p*
R^2	.393	Regression	5	8.68	$p<.05$
Adjusted *R* Square	.347	Residual	67		
Standard Error	6043.352				
Durbin Watson Test	8.683				

Hypothesis 3: There is no relationship between selected human resource variables and agricultural income/ha of arable land. The hypothesis was tested using multiple regression analysis.

Table 107 presents the inter-correlation matrix of the dependent variable, agricultural income per hectare of arable land (in U.S. $) and five independent variables: number of farmers per researcher, arable land per extension worker, percentage of subject matter specialists of total extension personnel, percentage of extension personnel with university degree or higher education, and number of farmers per extension worker. These five independent variables were free from multicollinearity. However, none of the independent variables were found to be significantly correlated with the dependent variable. All of these variables indicated a negative direction with agricultural income per hectare arable land.

Table 107: Inter-correlation Matrix Among Selected Human Resource Variables with Agricultural Income/ha Arable Land.

	Variable	X_1	X_2	X_3	X_4	X_5
X_1	No. of Farmers/Researcher	1.000				
X_2	Arable land/Extension worker	–.095				
X_3	% of Subject Matter Specialist	.004	.216			
X_4	Educational Level	–.345**	.248**	.411***		
X_5	No. of Farmers/Extension Worker	.370***	.093	.251	–.016	
Y	Agril. Income (U.S.$)/ha	–.042	–.042	–.083	–.042	–.048

Note— $N = 73$. $^{**}p<.01$, $^{***}p<.001$.

When this analysis was repeated for those low income countries having per capita GDP of less than U. S. $ 5,000, none of the variables were found significantly correlated with agricultural income. The variables: number of farmers per researcher ($r = –.1616$), arable land per extension worker ($r = –.2314$), percentage of extension personnel with university degree ($r = -.0819$), and number of farmers per extension worker ($r = -.1272$) indicated a negative direction with the dependent variable. The variable percentage of SMS ($r = .1537$) indicated positive direction but not significantly related.

To examine the influence of variables in the model, a multiple regression analysis was conducted with agricultural income as the dependent variable. The purpose of this analysis was to statistically examine the relationship between the dependent variable and the five independent variables in the model. Also, this analysis is used to estimate the magnitude of the relationship between the dependent variable and best linear combination of independent variables. It was, however, found that the Beta value (Table 108) of all independent variables are negative and not significant at the .05 level. This means that the dependent variable, agricultural income per hectare of arable land is not related with the independent variables in the model.

Table 108: Multiple Regression Coefficient with Agricultural Income /ha of Arable Land

	Variable	*B*	*T*	*Beta*	*Significance*
X_1.	No. of Farmers/Researcher	–.024	–.343	–.048	.7327
X_2.	Arable Land/Ext. Worker	–.007	–.193	–.024	.8479
X_3.	% of Subject Matter Specialist	–62.517	–.453	–.063	.6517
X_4.	Educational Level	–14.208	–.180	–.026	.8580
X_5.	No. of Farmers/Ext. Worker	–.051	–.095	–.012	.9243
	Constant	5126.261	1.526		.1318

When this analysis was repeated for low income countries having per capita GDP of less than U.S. $ 5,000, the variable, percentage of subject matter specialist, was found to have a significant relationship ($\beta = .333, p<.05$) with agricultural income/ha of arable land. The other independent variables in the model were not found to be significant at the .05 level of significance.

Table 109 indicates that only –.06 per cent of the variance in the agricultural income per hectare arable land was explained by the five independent variables and this relationship was not significant. Hence, the null hypothesis that there is no relationship between the selected human resource variables and agricultural income/ha of arable land is accepted.

When this analysis was repeated with the low income countries having per capita GDP of less than U.S $ 5,000, .06 per cent of the variance in agricultural income/ha arable land was explained by the variables in the model. However, the p =.1314, which was not significant (p>.05), hence, for low income countries, the null hypothesis was also accepted.

Table 109: The Overall F-Test

Multiple R	.100	Analysis of variance	$D.F.$	F	p
R^2	.010	Regression	5	0.137	p>.05
Adjusted R Square	–.063	Residual	67		
Standard Error	15718.030				
Durbin Watson Test	1.897				

Hypothesis 4: There is no relationship between selected human resources variables and total fertilizer consumption on kg/ha of arable permanent cropland. This hypothesis as tested by using multiple regression analysis.

Table 110 presents one tailed inter-correlation matrix of five independent variables with total fertilizer consumption. Of the five independent variables, percentage of subject matter specialists (r = .385***, p=.001), and percentage of extension personnel with a university degree or higher education show positive and higher (r = .352***, p =.001) levels of significance. This means that the higher percentage of both subject matter specialists and percentage of extension personnel with a university degree or higher education are related with higher levels of fertilizer consumption in kg/ha of arable and permanent cropland. The other three independent variables; number of farmers per researcher *(r = –.207),* arable land per extension worker (r = –.089*),* and number of farmers per extension worker (r = –.214*)* were not significantly related and indicate an inverse relationship, which would be expected.

Table 110: Inter-correlation Matrix Among Selected Human Resource Variables with Fertilizer Consumption in kg/ha Arable and Permanent Cropland

	Variable	X_1	X_2	X_3	X_4	X_5
X_1	No. of Farmers/Researcher	1.000				
X_2	Arable land/Extension Worker	–.048				
X_3	% of Subject Matter Specialist	–.090	.212			
X_4	Educational Level	–.266**	.270**	.419***		
X_5	No. of Farmers/Extension Worker	.536***	.082	.170	–.081	
Y	Fertilizer Consumption	–.207	–.089	.385***	.352***	–.214

Note— N = 80. **p < .01, ***p<.001.

When this analysis was repeated with the countries having per capita GDP of less than U.S. $ 5,000, none of the variables were found to be significantly correlated with the fertilizer consumption individually. The percentage of subject matter specialists (r = .1807) and percentage of extension personnel with university degree or higher education indicated a positive direction with total fertiliser consumption, while the variables number of farmers per researcher (r = –.2357), arable land per extension worker (r = –.2625), and number of farmers per extension worker (r = –.2427) indicated an inverse relationship as was the case in other set of analysis using the world-wide data set.

The purpose of this analysis was to identify which human resource variables are significantly related with the fertilizer consumption in kg/ha of arable and permanent cropland. To examine the influence of these independent variables in the model, a multiple regression analysis was conducted. Table 111 presents the standardised coefficient of five independent variables with respect to their influence on the dependent variable total fertilizer consumption in kg/ha arable and permanent cropland.

It is evident from Table 111 that the variables, arable land per extension worker, percentage of subject matter specialists, percentage of extension personnel with a university degree or higher education, and the number of farmers per extension worker were significant in explaining the fertilizer consumption in kg/ha of arable and permanent cropland at the .05 level of significance. Only the variable, number of farmers per researcher was not found to be significant in relation to fertilizer consumption.

Table 111: Multiple Regression Coefficient with Total Fertilizer Consumption in kg/ha Arable and Permanent Cropland

	Variable	*B*	*T*	*Beta*	*Significance*
X_1.	No. of Farmers/Researcher	2.52	0.10	.012	.9177
X_2.	Arable Land/Ext. Worker	–3.37	–2.07	–.211	.0417
X_3.	% of Subject Matter Specialist	3.17	3.36	.374	.0012
X_4.	Educational Level	1.08	2.08	.236	.0407
X_5.	No. of Farmers/Ext. Worker	–.00	–2.06	–.248	.0422
	(Constant)	33.20	1.54		.1256

When this analysis was repeated with the countries having per capita GDP of less than U.S. $ 5,000, only the percentage of subject matter specialists was found to be significant (β =. 2803, *p*<. 05). The other variables were not found to be significantly related at the .05 levels.

Table 112 indicates that 24.9 per cent of the variance in the total fertilizer consumption in kg/ha of arable and permanent cropland is explained by the variables; farmers per researcher, arable land per extension worker, percentage of subject matter specialists, percentage of extension personnel with university degree or higher education of total extension personnel, and number of farmers per extension worker. In other words, there is at least 95 per cent probability that 24.9 per cent of the variance in the total fertilizer consumption in kg/ha of arable and permanent cropland is due to five variables in the model. Hence, the null hypothesis that there is no relationship between the selected human resource variables and the total fertilizer consumption kg/ha of arable and permanent cropland is rejected.

When this analysis was repeated with the countries having per capita GDP of less than U.S.$ 5,000, the results indicated that 11.3 per cent variance in the total fertilizer consumption was explained by these variables in the model. The F level of significance was p =.0307, hence the null hypothesis was rejected.

Table 112: The Overall F-Test

Multiple *R*	.545	Analysis of variance	*D.F.*	*F*	*p*
R^2	.297	Regression	5	6.25	*p*<.05
Adjusted *R* Square	.249	Residual	74		
Standard Error	110.070				
Durbin Watson Test	1.678				

Financial Resources

The relationship of financial resources with agricultural development was tested separately. There were six financial resource variables: research expenditure per researcher, research expenditure per farmer, research expenditure per hectare arable land, extension expenditure per extension worker, extension expenditure per farmer, and extension expenditure per hectare land. Since all of these variables were computed from total research expenditure and total extension expenditure, these variables were found to be multicollinear after multicollinearity diagnosis. Hence, these variables were not included in the multiple regression analysis. Instead, a compound variable total research and extension (R&E) expenditure per farmer was computed adding the extension expenditure per farmer and research expenditure per farmer to estimate the relationship of total research (R) and extension (E) expenditure per farmer to the agricultural development indicators.

This analysis as indicated earlier, was conducted first for all the 114 responding countries and then for the countries which were having per capita GDP of less than U.S. $ 5,000. The same dependent and independent variables were used for both sets of analysis. The dependent variables were total cereal yield kg/ha, agricultural income per hectare arable land (U.S. $), farmer income (U.S. $), total fertilizer consumption (kg/ha), and the independent variable was total R & E expenditure per farmer. Four null hypotheses were tested in this section to assess the relationship of this financial resource variable with agricultural development indicators at .05 level of significance.

Hypothesis 1: There is no relationship between the financial resource variable, research and extension expenditure per farmer, and total cereal yield. Multiple regression analysis was used to test this hypothesis.

Table 113 presents one tailed inter-correlation matrix of the dependent variable, total cereal yield and the independent variable total R&E expenditure per farmer. The independent variable, total R&E expenditure per farmer (r = .5636***, p = <.001) was found to be significantly and positively correlated with total cereal yield. This positive direction of total R&E expenditure per farmer indicates that higher level of total research and extension expenditure per farmer is directly related to greater cereal yields. When this analysis was repeated for the countries having per capita GDP less than U. S. $ 5,000, the total R&E expenditure per farmer (r =.2077, p = >.20) was not found significant but indicate positive direction with total cereal yield. This result indicates that R&E expenditure per farmer is related with cereal yield for all responding countries and indicates a positive direction for the developing countries.

Table 113: Inter-correlation Matrix of Total R&E Expenditure per Farmer with Total Cereal Yield kg per Hectare

X_1.	Total R&E Expenditure per farmer	1.000
Y.	Total Cereal Yield, kg per hectare	.5636***

Note– N = 60. ***p<.001.

The purpose of the multiple regression analysis was to statistically examine the relationship of the dependent variable, total cereal yield with the total R&E expenditure per farmer as the financial independent variable in the model. This analysis estimates the magnitude of the relationship between the dependent variable and the independent variable. It has been widely observed that due to the financial problems, research and extension activities are not adequately carried out in many developing countries and many of these countries have chronic food shortage problem (malnutrition and under-nutrition); (see Appendix C, Table C-1). If it can be established that cereal yield is directly related with higher research and extension expenditure per farmer, then this finding would provide empirical support to the proposition that financial investment in agricultural research and extension systems in developing countries should be increased.

Table 114 shows the standardised coefficient of the independent variable, total R&E expenditure per farmer, with respect to its influence on the dependent variable total cereal yield kg/ha. It reveals that the variable total R&E expenditure per farmer, ($\beta = .5635$, $p<.001$) had a significant relationship with total cereal yield. This analysis, therefore, reveals that higher total R&E expenditure per farmer is directly related to total cereal yield. When this analysis was repeated to those countries having a per capita GDP of less then US $ 5,000, the variable, total R&E expenditure per farmer was found to be significant. The Beta value of total expenditure per farmer was $\beta = .4267$, $p<.01$.

Table 114: Multiple Regression Coefficient with Total Cereal Yield as Dependent Variable

	Variable	*B*	*T*	*Beta*	*Significance*
X_1	Total R&E Expenditure/Farmer	4.3618	5.196	.5635	.0000
	(Constant)	1708.0763	11.430		.0000

Table 115 indicates that for the worldwide data set, 30.5 per cent of the variance in the total cereal yield was explained by total R&E expenditure per farmer. The result of the F-Test shows that more than 30 per cent of the variance explained by this variable is not by chance. The level of significance, $p<.0001$ indicates that there is more than a 99.9 per cent probability that 30.5 per cent of the variance in the total cereal yield was due to the total R&E expenditure per farmer. Hence, the null hypothesis, that there is no relationship between the financial resources variable and total cereal yield, was rejected. When this test was repeated with the countries having GDP of less than US $ 5,000, the comparable statistics were as follows:

Table 115: The Overall F-Test

Multiple *R*	.5635	Analysis of variance	*D.F.*	*F*	*p*
R^2	.3176	Regression	1	26.99	p<.0001
Adjusted *R* Square	.3058	Residual	58		
Standard Error	1060.3964				

Multiple *R* was .4094, R^2 =.1676, adjusted *R* Square = .15099, Standard Error = 915.53, *F* =10.0702, and *p* = .0026, which was <.05. Hence, the null hypothesis was rejected at the .05 level of significance.

Hypothesis 2: There is no relationship between financial resources variable, research and extension expenditure per farmer, and farmer income (U.S.$). It was tested by multiple regression analysis.

Table 116 presents a one tailed inter-correlation matrix of the independent variable, total R&E expenditure per farmer, with farmer income as the dependent variable. The independent variable (r= .6930***, p = <.001) is very significantly and positively correlated with farmer income. This result indicates that if the total R&E expenditure per farmer is increased, then farmer income can also be expected to increase. When this analysis was repeated for those countries having a per capita GDP of less than U. S. $ 5,000, this variable was found to be significantly related with farmer income. Therefore, the variable, total R&E expenditure per farmer is significantly and positively correlated with farmer income in both data sets.

Table 116: Inter-correlation Matrix of Total R&E Expenditure per Farmer with Farmer Income

X_1	Total R&E Expenditure per Farmer	1.000
Y	Farmer Income	.4267***

Note— *N* = 52. ***p<.001.

The purpose of this multiple regression analysis was to identify whether or not the financial resource variable is significantly related with farmer income. To examine the influence of the independent variables, total R&E expenditure per farmer on farmer income as dependent variable, a multiple regression analysis was conducted.

Table 117 presents the standardised coefficient of total R&E expenditure per farmer with respect to its influence on the dependent variable, farmer income. It is evident from this table that the variable total R&E expenditure per farmer (β =.6929, p<.0001) is highly significant in explaining the farmer income. It can be inferred from this result that increased farmer income is directly related to increased research and extension expenditure per farmer. When this analysis was repeated with those countries having per capita GDP of less than U.S. $ 5,000, total expenditure per farmer was found significant at the .05 level of significance. The Beta value of total R&E expenditure per farmer was (β =.4267, *p*<.01).

Table 117: Multiple Regression Coefficient with Farmer Income as Dependent Variable

	Variable	*B*	*T*	*Beta*	*Significance*
X_1	Total R&E Expenditure/ Farmer	25.8484	6.796	.6929	.0000
	(Constant)	719.281	.962	.3405	

Table 118 indicates that 46.9 per cent of variance in the farmer income is explained by the independent variable, total R&E expenditure per farmer. The level of significance (*p*<.0001) indicates that there is more than 99.9 per cent probability that 46.9 per cent of the variance in the farmer income is due to the total R&E expenditure per farmer. Hence, the null hypothesis is rejected. When this analysis was repeated with the countries having a per capita GDP of less than U.S. $ 5,000, the multiple R = .4267, R^2 = .1820, Adjusted R Square =.1630, Standard error = 901.966, F = 9.5723, and F significance =<.01, which was p<.05. Hence, the null hypothesis was also rejected for the developing country data set.

Table 118: The Overall F-Test

Multiple R	.6929	Analysis of *Variance*	*DF*	*F*	*p*
R^2	.4802	Regression	1	46.19	*p*<.0001
Adjusted R Square	.4698	Residual	50		
Standard Error	4951.5573				

Hypothesis 3: There is no relationship between financial resources variable; research and extension expenditure per farmer, and agricultural/ha of arable land. It was tested by multiple regression analysis.

Table 119 presents a one tailed inter-correlation matrix of dependent variable, agricultural income per hectare arable land in U.S. $, and the independent variable total R&E expenditure per farmer. The independent variable total R&E expenditure per farmer (*r* = .4515***, *p* = <.01) was found to be significant and positively related with agricultural income per hectare arable land. The positive direction of this relationship indicates that the higher the total R&E expenditure per farmer is directly related to higher agricultural income/ha of arable land. When this analysis was repeated for those countries having a per capita GDP of less than U. S. $ 5,000, the total R&E expenditure per farmer (r = .4724***, p = <.001) was found to be significant and positively correlated with agricultural income.

The focus of multiple regression analysis was to statistically examine the possible relationship between the dependent variable, agricultural income/ha of arable land, with the independent variable, total R&E expenditure per farmer. The other purpose of this analysis was to estimate the magnitude of the relationship between the dependent and independent variables.

Table 119: Inter-correlation Matrix of Total R&E Expenditure per Farmer with Agricultural Income

		X_1
X_1	Total R&E Expenditure per farmer	1.000
Y	Agricultural Income	.4515***

Note—$N = 52$. ***$p<.001$.

Table 120 presents the standardised coefficient of total R&E expenditure per farmer with respect to its influence on agricultural income (U.S $)/ha of arable land . It reveals that the variable total R&E expenditure per farmer (β = .4515, $p<.0008$) is significant at the .05 level of significance. When this analysis was conducted with countries having a per capita GDP of less than U. S. $ 5,000, total R&E expenditure per farmer (β=.4723, $p>.0011$) was found to be significant at the .05 level of significance. The Beta value of total R&E expenditure per farmer indicates positive direction in both sets of analysis.

Table 120: Multiple Regression Coefficient with Agricultural Income as Dependent Variable

	Variable	*B*	*T*	*Beta*	*Significance*
X_1	Total R&E Expenditure/ Farmer	6.7962	3.578	.4515	.0000
	(Constant)	686.7624	1.840	.0717	

Table 121 indicates that 18.8 per cent of variance in the agricultural income/hectare arable land was explained by total expenditure per farmer. The level of significance ($p<.0001$) indicates that there are 99 per cent probability that 18.8 per cent of the variance in the agriculture income is due to the total R&E expenditure per farmer. Hence, the null hypothesis is rejected. When this analysis was repeated with the countries having a per capita GDP of less than U.S. $ 5,000, the multiple $R = .4723$, $R^2 = .2231$, Adjusted $R^2 = .2050$, Standard error=1118.2592, $F = 12.35$, and F significance =.0011, which was $p<.01$. Hence, the null hypothesis was rejected in both sets of analysis.

Table 121: The Overall F-Test

			DF	*F*	*p*
Multiple *R*	.4515	Analysis of Variance			
R^2	.2039	Regression	1	12.80	$p<.001$
Adjusted *R* Square	.1879	Residual	50		
Standard Error	2472.1016				

Hypothesis 4: There is no relationship between financial resources variable, research and extension expenditure per farmer, and total fertilizer consumption kg/ha of arable and permanent cropland. It was tested by multiple regression analysis.

Table 122 presents a one tailed inter-correlation matrix of the independent variable, total R&E expenditure per farmer with total fertilizer consumption. The independent variable, total R&E expenditure per farmer, (r = .6688***, p = <.001) was found to be significant and positively correlated with total fertilizer consumption. The result indicates that an increase of total R&E expenditure per farmer is directly related to higher levels of fertilizer consumption. When this analysis was repeated for developing countries (per capita GDP of less than U.S. $ 5,000), the variable, total R&E expenditure per farmer (r = .4294***, p = <.001) was found to be positively and significantly related to fertilizer consumption in kg/ha of arable and permanent cropland. Therefore, it may be concluded that increased research and extension expenditure per farmer is related to higher levels of fertilizer consumption.

Table 122: Inter-correlation Matrix of Total R&E Expenditure per Farmer with Total Fertilizer Consumption kg/ha Arable and Permanent Cropland

X_1	Total R&E Expenditure per Farmer	1.000
Y	Total Fertilizer Consumption/Farmer kg/ha	.6688***

Note– *N*= 61. ***$p<.001$.

The relationship between these two variables was examined through multiple regression analysis. Table 123 presents the standardised coefficient of the independent variable, total R&E expenditure per farmer, with respect to its influence on total fertilizer consumption in kg/ha of arable and permanent cropland. It reveals that the variable, total R&E expenditure per farmer (β =.6688, p<.0001) had a very significant relationship with total fertilizer consumption. When this analysis was repeated for the developing countries (per capita GDP of less than U. S. $ 5,000), the Beta value of total R&E expenditure per farmer (β = .4293, p<.01) indicates a significant relationship with total fertilizer consumption at the .05 level of significance.

Table 123: Multiple Regression Coefficient with Total Fertilizer Consumption

	Variable	*B*	*T*	*Beta*	*Significance*
X.	Total R&E Expenditure/Farmer	5.033	7.715	.883	.0000
	(Constant)	2.7068	.087		.9311

Table 124 indicates that 43.7 per cent of variance in the total fertilizer consumption was explained by total R&E expenditure per farmer. The level of significance, (***p<.0001) indicates that the result is very significant. In otherwords, there is 99.9 percent probability that 43.7 per cent of the variance in the total fertilizer consumption is due to the total R&E expenditure per farmer. Hence, the null hypothesis that there is no relation between financial resource variable and total fertilizer consumption in kg/ha of arable and permanent cropland is rejected.

When this analysis was repeated for the developing countries (per capita GDP of less than U.S. $ 5,000), the following statistics were computed: the multiple R=.4243, R^2= .1843, Adjusted R Square =.1680, Standard error = 68.8643, F = 11.302, and F significance =.0015, which was p<.01. Hence, the null hypothesis was rejected. In other words, the analysis found that null hypothesis was rejected in both sets, which means the total research and extension expenditure per farmer is related to fertilizer consumption.

Table 124: The Overall F-Test

Multiple R	.6688	Analysis of Variance	*DF*	*F*	*p*
R^2	.4473	Regression	1	47.76	p<.0001
Adjusted R Square	.4379	Residual	59		
Standard Error	275.1466				

5

SUMMARY, CONCLUSIONS, DISCUSSION AND RECOMMENDATIONS

Summary

Agricultural research and extension are two major systems engaged in agricultural technology development and transfer aimed at agricultural and rural development. Currently these two systems have been established in nearly every country of the world. Worldwide, these two systems involve annual expenditures of an estimated 10 billion dollars and they employ nearly 1 million people. Specific information on the current status of the human and financial resources being invested in these two systems, and their relationship with and contribution to agricultural development is not fully known. Therefore, the main focus of this investigation was to study the human and financial resources being invested in approximately 114 national agricultural research and extension systems worldwide and to determine their relationship with selected agricultural development indicators; total cereal yield, farmer income, agricultural income, and total fertilizer consumption.

The goals of this study were four fold. The first goal was to investigate the present human resource situation of approximately 114 national agricultural research and extension worldwide and their relationship with agricultural development indicators. The second goal was to investigate the quality of extension personnel in terms of their educational level and its relationship with the agricultural development indicators. The third goal was to investigate the level of financial resources being invested in these agricultural research and extension systems and to determine their relationship with agricultural development indicators. Finally, the fourth goal was to assess the overall relationship of these human and financial resource measures, as independent variables, with the agricultural development indicators, as dependent variables.

For each goal, specific objectives were pursued. Under objective one, the relationship of seven human resource variables, including the number of farmers per extension worker, arable land per extension worker, number of farmers per researcher, arable land per researcher, number of extension worker per researcher, percentage of SMS of total extension personnel, and percentage of female extension personnel of total extension personnel were studied. The relationship of these seven independent variables were tested with four dependent variables using twenty-eight hypotheses and Chi-Square analysis. Under objective two, the percentage of extension personnel with university degree or higher qualification was used as the independent variable, in relation to the four agricultural development indicators as dependent variables. Four hypotheses were tested in this objective. Under objective three, research expenditure per farmer and extension expenditure per farmer were used as the independent variables. Eight hypotheses were tested under this objective using Chi-Square analysis. Under objective four, five human resource variables were used collectively to assess their relationship with the four dependent variables, using multiple regression analysis. Then two financial resources variables were used to collectively assess their relationship with these same four dependent variables. Under objective four, a total of eight hypotheses were tested. Overall, a total of 48 hypotheses were tested and information is presented in 124 tables. In addition to this multiplicative effect of these hypotheses was further tested using one-way analysis of variance (ANOVA) of the log values of these independent variables.

The extension data for the study were collected from 114 countries of the world. The list of the responding countries is included in Appendix B. By region, 31 countries were from Africa, 26 from Asia, 14 from Europe, 24 from Latin America, 3 from North America, and 16 from the Near East. While categorizing by economic group, 37 countries were considered to be low income countries, 41 were lower middle income, 11 were upper middle income, and 25 were categorized as high income countries. By agricultural population group, 32 countries were classified as low agricultural population countries, 26 were lower middle agricultural population countries, 20 were upper middle agricultural population countries and 36 were categorized as higher agricultural population countries. Data on each variable were aggregated by geographic region and economic group throughout objectives one through three.

The human and financial resource data on national agricultural extension systems were collected by a mail questionnaire through the auspices of the Food and Agricultural Organization (FAO) of the United Nations. The questionnaire were developed by a team of researchers including the investigator at the University of Illinois. A copy of the questionnaire is enclosed in Appendix A. The human and financial resources data of agricultural research systems were taken from the ISNAR Agricultural Research Indicator Series. The Agricultural Development indicator variables, total cereal yield and total fertilizer consumption, were taken from the FAO Production Yearbook and FAO Fertilizer Yearbook, respectively. The reaming two agricultural development indicators, including farmer income and agricultural income were computed from secondary data published by the World Bank (World Bank 1990-91).

Two sets of analysis were conducted under each of the four objectives. The first set of analyses included 114 responding countries and this is referred to as the worldwide data set. The second set of analyses included only those developing countries having per capita GDP of less than U.S. $ 5,000. The rationale for this second set of analyses was to find out whether the findings of the first set hold the same result in the second set of analysis. The data were analysed using the SPSS statistical package for the microcomputer. Statistical methods used in the study were descriptive statistics, Chi-Square, Multiple regression, and ANOVA of log analysis. Under descriptive statistics, data on the different variables were compared across geographic regions and economic groups. Then Chi-Square tests were performed to assess the expected relationship of the various dependent and independent variables; four hypotheses were tested in each case. In order to assess the collective relationship of different groups of independent variables with each dependent variable, a multiple regression analysis was used. Again, the analysis was completed for both data sets. Four hypotheses were tested with the human resources variables and four hypotheses were tested with the financial resources variables. The significant findings of the study are summarized below.

Summary of the Findings

1. Descriptive analysis found that there are 542,566 extension personnel and 119,226 agricultural researchers serving approximately 4.2 billion population in 114 countries. Out of

this total population, about 990.616 million people are economically active in agriculture. The number of farmers per researcher ranges from 216 in North America to 30,144 in Africa. The number of farmers per extension worker ranges from 233 in North America to 2,087 in Asia. The worldwide average of the number of farmers per researcher was 8,308 and the number of farmers per extension worker was found to be 1,826.

Chi-Square analysis found that fewer farmers per extension worker and per researcher are significantly related with the agricultural development indicators. In other words, fewer number of farmers per extension worker and per researcher or the greater the intensity of research and extension activities is related to increased cereal yields, farmer income, agricultural income, and fertilizer consumption in both sets of data, with the following exceptions. The number of farmers per extension worker and per researcher were not found to be significantly related to agricultural income in the developing country data set.

2. Descriptive analyses found that agricultural researchers and extension workers in these 114 countries have to cover more than 963.35 million hectares of arable land. The arable land per extension worker ranges from 1,032 hectare in Asia to 15,411 hectares in North America, while the arable land per researcher ranged from 3,603 hectares in Europe to 31,874 hectares in Africa. The worldwide average was 1,776 hectares of arable land per extension worker and 8,080 hectares per researcher.

   Chi-Square analysis found that the fewer number of hectares of arable land per extension worker was related to increased agricultural income and fertilizer consumption in both data sets. Arable land per extension worker was not found to be significant for cereal yield and farmer income in both data sets. Fewer number of hectares of arable land per researcher was found to be significantly related to increased cereal yields, farmer income, agricultural income, and fertilizer consumption in both sets of data.

3. Descriptive analysis found that the ratio of extension workers per researcher ranged from 1/1 in Europe and North America, to 18/1 in Africa. The same pattern was found in

the various economic groups, with a 1/1 ratio in high income group up to an 8/1 ratio among low income countries. The worldwide average was 4.5 extension workers per researcher.

Chi-Square analysis found that fewer numbers of extension workers per researcher was significantly related to increases in farmer income and fertiiizer consumption in both data sets. This relationship was also significant for cereal yield in the worldwide data set, but not significantly related in developing countries data set. The relationship between agricultural income and the extension worker-researcher ratio was not found to be significant in either data set.

4. Descriptive analysis found that the percentage of SMS ranged from 11.6 per cent in Latin America to 45.2 per cent in North America. The low income countries have a low percentage of SMS (10.6 per cent) and high income countries have highest per cent (25 per cent). The worldwide average was found to be 14.4 per cent. Chi-Square analysis indicated that percentage of SMS is not significantly related, individually, with any of the agricultural development indicators. However, as noted below, the SMS variable was found to be significant factor in the multiple regression and ANOVA analysis.
5. Descriptive analysis found that the estimated percentage of females who are economically active population in agriculture was the highest in Africa (39 per cent) and the lowest in Latin America (12.6 per cent). More than 26 per cent of the economically active population in agriculture are women worldwide. The percentage of female extension workers of total extension workers on the other hand was found to be lowest in Africa (8.2 per cent) and highest in North America (30.7 per cent). The worldwide average of female extension personnel was found to be only 11.1 per cent of total extension personnel. Chi-Square analysis found that percentage of female extension personnel is not significantly related, individually, to the agricultural development indicators in any of the data sets.
6. Descriptive analysis found that the percentage of extension personnel with a university degree or higher educational qualification was much lower in developing countries —

Africa (11.8 per cent), Asia (23.2 per cent), and Latin America (31.3 per cent) in comparison to North America (90.3 per cent). Chi-Square analysis found that percentage of extension personnel with at least a university degree was significantly and positively related to cereal yields in both the data sets. Also, the educational level of extension personnel was found to be significantly related to farmer income and fertilizer consumption in the worldwide data set. However, this variable was not significantly related to agricultural income in both of data or to farmer income and fertilizer consumption in the developing country data set.

7. Descriptive analysis found that research expenditure in North America was the highest of any region (U.S. $ 1,760.9 Million), while Africa was the lowest (U.S. $ 204.68 Million). Research expenditure for the 114 countries included in this analysis was estimated at U.S. $ 5.4 billion. Research expenditure per farmer was found to be lowest in Africa (U.S. $ 2.040 and highest in North America (U.S. $ 498.70). Worldwide research expenditure per farmer was very significantly and positively related to cereal yield, farmer income, agricultural income and fertilizer consumption in the worldwide data set. It was also significantly related to farmer income and fertilizer consumption in the developing countries data set. It was however, not found to be significant in its relationship with cereal yield and agricultural income in the developing countries data set.

8. The overall level of extension expenditure was found to be lowest in Africa and the highest in the North America, the situation was the same with research expenditures. Extension expenditure per farmer was found to be the lowest in Asia (U.S. $ 1.32) followed by Africa (U.S. $ 1.87) and the Near East (U.S.$ 5.68). Extension expenditure per farmer categorized by economic groups, ranged from U.S. $ 0.71 in low income countries to U.S. $ 58.48 in high income countries. Worldwide average extension expenditure per farmer was found to be U.S. $ 2.34.

Chi-Square analysis found that extension expenditure per farmer was significantly and positively related to cereal yield, farmer income, agricultural income, and fertilizer consumption in the worldwide data set. Also, it was found to be significantly related to farmer income and fertilizer consumption in the developing countries. However, it was not found to be significantly related to cereal yield and agricultural income in developing countries data set.

9. Multiple regression analysis found that the number of farmers per researcher, arable land per extension worker, percentage of subject matter specialists, percentage of extension personnel with a university degree or higher education, and number of farmers per extension worker are collectively very significant in their relationship with the dependent variable, total cereal yield, in both data sets. However, none of these variables were found to be significantly related to cereal yield, on an individual basis, in the developing country data set; but the variables, arable land per extension worker, percentage of subject matter specialists, and percentage of extension personnel with university degree or higher education were approaching significance. In the worldwide data set, the percentage of extension personnel with university degree or higher education and number of farmers per extension worker were found to be significantly related, on an individual basis, with cereal yields and the percentage of SMS and arable land per extension workers were approaching significance at .05 and .07 levels respectively. The amount of variance explained by these regression models was 23.8 per cent for the worldwide data set and 11.4 per cent for the developing country data set.

According to the Beta values obtained from multiple regression analysis of variance of five independent human resource variables with total cereal yield as the dependent variables, the order of influence was as follows:

(a) The percentage of extension personnel with at least a university degree was the most important variable in helping explain changes in total cereal yield in the worldwide data set, while fewer hectares of arable land per extension worker was the most important variable influencing total cereal yield in the developing country data set. In other words, the higher the percentage of university level educated extension personnel and the fewer the hectares of arable land per extension worker, the more likely total cereal yield would be higher in all responding countries and in the developing countries, respectively.

(b) The second most important variable in influencing total cereal yield was fewer number of farmers per extension worker in the worldwide data set and the percentage of SMSs in the developing country data set.

(c) The third most important variable influencing total cereal yield was the percentage of SMSs in the worldwide data set and the percentage of extension personnel with university degree or higher educated for the developing countries data set.

(d) The fourth most important variable in helping explain total cereal yield was fewer hectares of arable land per extension worker in the worldwide data set and fewer number of farmers per extension worker for the developing country data set.

(e) Finally, the fifth most important variable in relation to total cereal yield was fewer number of farmers per researcher in both data sets.

10. Multiple regression analysis of variance found that the five independent human resource variables are collectively very significant in their relationship with farmer income, as the dependent variable, in both data sets. Out of these five variables, fewer hectares of arable land per extension worker and a higher percentage of university educated extension personnel were found to be significantly and positively related to the dependent variable for the developing country data set. In the worldwide data set, with the exception of the number of farmers per researcher, the remaining four variables were found to be significantly related to farmer income. The amount of variance explained by these variables for the worldwide data set was 34.79 per cent and 16.78 per cent for the developing countries. The order of influence, as indicated by the Beta value are the following:

(a) The percentage of SMS of total extension personnel was

the most important variable in helping explain increases in farmer income in the worldwide data set and fewer number of farmers per extension worker was the most important variable in relation to farmer income in the developing country data set. In other words, a higher percentage of SMS, and a fewer number of farmers per extension worker were the most important factors in explaining increases in farmer income in all responding countries and in the developing countries respectively.

(b) The second most important variable in explaining increases in farmer income was the percentage of extension personnel with a university or higher degree in both data sets.

(c) The third most important variable in relation to farmer income was more hectares of arable land per extension worker in the worldwide data set and fewer hectares of arable land per extension worker for the developing countries. This contradiction was due to differences in farm size in the industrially developed countries, as contrasted with the developing countries.

(d) The fourth most important variable in explaining farmer income was fewer number of farmers per extension worker in the worldwide data set and fewer number of farmers per researcher in the developing countries.

(e) Finally, the fifth most important variable in helping explain farmer income was fewer number of farmers per researcher in the worldwide data set and the percentage of SMS for developing country data set.

11. Through multiple regression analysis of variance it was found that the five independent human resources variables were not significant in explaining agricultural income as the dependent variable. Only the percentage of subject-matter specialists was found to be significantly related to agricultural income in the developing countries, but even this variable was not useful in helping much of the variance in the dependent variables.

12. Through multiple regression analysis of variance, the five independent human resources variables were, collectively, found to be significant in explaining differences in fertilizer

consumption as the dependent variable. However, out of these five variables in the model, the percentage of SMS was the only variable that was individually significant in the developing country data set. Except for the number of farmers per researcher, in the worldwide data set all four other variables were found to be very significant in explaining differences in total fertilizer consumption. The amount of variance explained in this model was 24.9 per cent and 11.13 per cent for all responding countries and developing countries respectively. The order of influence of the independent variables, as indicated by the Beta values are listed below:

(a) Increase in the percentage of SMS of total extension personnel was the most important variable in helping explain increases in total fertilizer consumption in both data sets.
(b) The second most important variable in explaining total fertilizer consumption was fewer number of farmers per extension worker in the worldwide data set and fewer number of farmers per researcher for the developing country data set.
(c) The third most important variable in explaining total fertilizer was the percentage of extension personnel with a university or higher level degree in the worldwide data set and fewer number of farmers per extension worker in the developing countries.
(d) The fourth most important variable in explaining total fertilizer consumption was fewer number of farmers per extension worker in the worldwide data set and fewer number of farmers per researcher for the developing country data set.
(e) Finally, the fifth most important variable was the number of farmers per researcher in the worldwide data set and the percentage of extension personnel with a university or higher degree in the developing country data set.

13. Through multiple regression analysis, the independent financial resource variable total expenditure per farmer was

found to be significant in its relationship with total cereal yield in both data sets. The amount of variance explained by this variable for all countries and for the developing countries was 30.5 and 15.0 per cent respectively. The independent variable total expenditure per farmer was found to be very significant in its relationship with the dependent variable in both data sets.

14. Multiple regression analysis of the independent financial resource variable total expenditure per farmer was also found to very significant in its relationship to farmer income. The amount of variance explained by this variable for the worldwide and developing country data sets was 46.9 per cent and 16.3 per cent respectively. This variable was found to be significantly related in both data sets.
15. Through multiple regression analysis, the independent financial resource variable total expenditure per farmer was found to be very significant in its relationship with agricultural income. The variance explained by this variable was 18.7 per cent and 20.5 per cent for all responding countries and for the developing countries respectively. In other words, total expenditure per farmer was found to be significant for both data sets.
16. Through multiple regression analysis, the independent financial resource variable total expenditure per farmer was found to be significantly related to total fertilizer consumption. The amount of variance explained by this variable was 43.7 per cent and 16.8 per cent for all countries and for the developing countries respectively. Total expenditure per farmer was found to be very significant in its relationship to fertilizer consumption in both data sets.

Conclusion

Based on the analysis of two sets of data, the following conclusions may be drawn:

1. The lack of adequate numbers (quantity) of agricultural research and extension personnel in the developing countries, which has been mentioned by numerous scholars, was empirically documented by this study. In addition, the level of human resources available for agricultural research

and extension is significantly related to agricultural development, as measured by selected indicators.

2. The number of farmers per extension worker is comparatively very high in developing countries in contrast with fewer number of farmers per extension worker in Europe and North America. Fewer number of farmers per extension worker, or increased extension intensity, is very significantly related to increased agricultural development as measured by all four indicators.
3. The amount (hectares) of arable land per extension worker was comparatively much less in developing countries in contrast with the industrialized countries of North America and Europe. Fewer number of hectares of arable land per extension worker was found to be significantly related to increased farmer income and fertilizer consumption in both data sets.
4. Also, the fewer number of farmers per researcher was to be significantly related to the agricultural development indicators, but developing countries have higher farmer-researcher ratios, such as 30,144 to 1 in Africa, in comparison to 216 farmers per researcher in North America.
5. As noted by extension scholars, the educational level of extension personnel in most developing countries was found to be grossly inadequate. The vast majority of extension personnel in developing countries have less than a university level education and many have only secondary school diplomas or higher education was positively and significantly related with agricultural development indicators.
6. The percentage of female extension personnel of total extension personnel appears to be inadequate, not only in developing countries but also throughout the world (11.1 per cent of all extension personnel worldwide are women) in that more than 26 per cent of all farmers worldwide are female. It seems reasonable to conclude that many female farmers, especially in many developing countries may not be getting appropriate information from the available extension personnel.
7. The percentage of subject-matter specialists of total extension personnel as noted by extension scholars was

found to be about 14.4 per cent worldwide. A higher percentage of subject-matter specialists was found to be very significantly related to increased agricultural development.

8. The extension worker-researcher ratio in Europe and North America was found to be about 1/1, in contrast with an 18/1 ratio in Africa. A more balanced extension worker-ratio was found to be positively and significantly related to agricultural development.
9. Extension expenditure in developing countries was found to be very low in comparison to the industrially developed countries. It was found that total expenditure per farmer was significantly related to cereal yield, farmer income, agricultural income, and fertilizer consumption in both data sets.
10. Research expenditure per farmer in the developing countries is also much lower when compared with North America.
11. Five human resources variables were found to be significantly related to cereal yield, farmer income, and fertilizer consumption in both data sets collectively.
12. Total research and extension expenditure per farmer was found to be related to all four dependent variables in both data sets.

Discussion of the Findings

The central focus of this research was to determine if increased investment (human and financial resources) in agricultural research and extension has a direct, measurable impact on agricultural development. In addition, it was concerned with learning about the current level of investment in agricultural extension and research systems, what is the relative quality of extension personnel worldwide, how many female extension workers are currently employed in different regions of the world, what is the percentage of SMS to the total extension personnel, what is the current level of research and extension expenditure per farmer, and what relationship do these human and financial resource factors have with selected agricultural development variables. This research is the first of its kind to attempt to answer many of these unsolved questions through a comprehensive, worldwide study.

Based on this analysis, it appears that investment in extension has a direct impact on agricultural productivity and development, contrary to the conventional wisdom and practice in some donor agencies. In fact, it appears that increased investment in extension merits equal, if not higher importance, than further investment in research. However, it must be emphasized that it is not an either or question, but that both agricultural research and extension in developing countries require increased levels of human and financial resources investment if agricultural productivity and food security are to be increased.

There are other important implications of this study that can be used to enhance the effectiveness of agricultural research and extension systems and their impact on agricultural development. For example, the study found that the highest percentage of responding countries (31.6 per cent) are still predominantly agricultural economies. The majority of the population in these countries basically depend on agriculture for their livelihood. Hence, the agricultural development of these countries becomes an essential endeavour. Economically, these countries are in the lowest per capita income group of countries. This finding supports the general assumption that a higher percentage of population in agriculture is related to lower per capita income. Also, these poor countries are characterized by low literacy rates and the low use of radio and television to communicate agricultural information to farmers.

Because of these factors, agriculture in these countries tends to be, as Prof. Theodore Schultz said, traditional agriculture. In other words, farmers follow traditional rather than profit oriented cultivation practices. Also, agricultural research and extension systems in these countries tend to be inadequate. Farmers in such countries lack of technical knowledge about improved agricultural production practices including such things as timely planting and irrigation, and the use of improved crop varieties, agro-chemicals, and fertilizer. Dissemination of knowledge and information about these improved practices will have paramount effect on agricultural development in general and cereal yield in particular.

To disseminate the latest research findings to farmers, adequate number of extension personnel are required. The findings of this study empirically support the proposition that fewer number of farmers per extension worker is directly related to cereal yield, farmer income, and fertilizer consumption. ANOVA of log analysis also verified this assumption that if the number of farmers per extension workers decreases, then cereal yield, farmer income, and fertilizer consumption will increase. This empirical finding should be used to convince the policy and decision makers to increase the number and quality of researcher and extension workers.

The main role of SMSs is to update field extension workers with current knowledge through in-service training and back stopping so that the latest technology can reach all groups of farmers. However, it was found that the percentage of SMS is comparatively very low in many countries of Africa, Asia, Latin America, and the Near East. Although, percentage of SMSs was not found to be significantly related to the dependent variables in the multiple regression and ANOVA of log analysis for farmer income and fertilizer consumption. Therefore, an increased percentage of SMSs appears to be empirically justified in most developing countries, however, this relationship needs further investigation in future studies.

Mao Tse-Tung said "Women hold up half the sky". A high percentage of the farmers in many developing countries are women because they are basically uneducated or undereducated due to gender discrimination. They are generally less qualified for white collar jobs in urban areas than their male counterparts, so many females remain in agriculture. To educate these women farmers about the latest agricultural technology and thereby increase their productivity and incomes, the number of female extension worker in these countries needs to be increased. Some scholars mentioned that there is no problem in female farmers seeking advice from male extension workers, but this situation is heavily influenced by the socio-cultural situation.

For example, in most of the Asian countries, female dairy farmers will seldom come forward to ask about matters related with pregnancy, artificial insemination and other reproduction related matters with male extension workers. In this situation, women farmers would feel more comfortable with female extension workers. Also, it has been experienced that female extension personnel are more convincing to both male and female farmers. Therefore, the number of female extension workers needs to be increased extensively in countries where women play a major role in agricultural production.

Although in Chi-Square analysis, the proportion of female extension personnel were not found to be significantly related to the agricultural development indicators, the ANOVA of log analysis indicated that a higher percentage of female extension personnel was significantly related to cereal yield and fertilizer consumption for worldwide data set, and its relationship with farmer income was approaching significance (>.07). For the developing countries data set, the percentage of female extension personnel was significantly related to fertilizer consumption. In order to do a better job of educating female farmers worldwide, the percentage of female extension personnel should be increased.

The professional competency of extension workers in developing countries has been questioned by many extension scholars. This hypothesis was tested in this study and the educational qualification of the extension personnel was found to be directly related with selected agricultural development indicators. Although, this finding needs further study, the role of extension workers is similar to that of agricultural teachers. If agricultural teachers are lacking in their technical knowledge when teaching students, then they will be constrained in their activity to teach and communicate knowledge about improved agricultural practices.

The same situation is applicable for agricultural researchers and extension educators. If they are inadequately trained, they will be constrained in their ability to carry out their agricultural research and extension responsibilities, with result agricultural development will be hindered. This study found that percentage of extension personnel with a university degree or higher education is much lower in Africa, Asia, and Latin America than Europe and North America. In order to strengthen the agricultural research and extension systems in developing countries, the educational level of these research and extension personnel needs to be increased.

It has been pointed out by numerous scholars that due to the lack of finance, agricultural research and extension systems are not performing as they are expected. Also, it was pointed out that the most effective researchers and extension workers frequently resign their position in order to get better paying position in the private sector. Also, due to the inadequacy of recurrent funds, the mobility of extension workers and other programme activities are also restricted, which in turn affects the frequency and type of extension contact with the farmers. This study has provided strong empirical support to the proposition that increased expenditure in research and extension has an important impact on agricultural development. Hence, to provide the adequate transportation, housing, and other programme support resources needed by research and extension workers, increased allocation of financial resources to research and extension is essential to improving the performance of these systems.

Recommendations

1. The study found that fewer number of farmers per extension worker and per researcher are significantly related with agricultural development. Hence, the number of extension

worker and researchers should be increased, particularly in Africa, Asia, Latin America, and the Near East. In other words, the intensity of research and extension services to farmers should be strengthened by employing more research and extension personnel to achieve the agricultural development objectives in these developing countries.

2. The quality (educational level) of extension personnel was directly related with agricultural development, therefore, efforts need to be made to increase the qualifications and skills of extension personnel in developing countries through in-service education and training programme.
3. Research and extension expenditure per farmer was found to be significantly related to agricultural development but the current level of investment is inadequate in most developing countries. Therefore, research and extension investment needs to be increased to enhance the development and transfer of new agricultural technology. Increased expenditure is needed to provide adequate transportation, housing, programme costs, and remunerative salaries. By providing the adequate salaries, competent research and extension personnel can be recruited and retained.

Suggestions for Further Investigations

1. The percentage of subject-matter specialists to total extension personnel was not found to be significantly related in Chi-Square analysis, but it was very significant in the multiple regression analysis of the agricultural development indicators included in this study. This finding requires further analysis at local, regional, and global level.
2. Based on the current literature, it is assumed that females are largely bypassed by male extension personnel. However, in the Chi-Square analysis, this variable was not found to be significantly related with any of the agricultural development variables in this study. In fact, the issue is one of women farmers to improve agricultural technology and other extension services. This issue requires further study at national, regional and global level to determine the most effective means of reaching women farmers with improved technology.
3. Projections of the human and financial resources required by

agricultural research and extension systems is another important area that merits further study.

4. A time series study of the human and financial resources being invested in agricultural research and extension is required to measure the specific impact of these investments on agricultural development.

APPENDIX A

QUESTIONNAIRE PART 1

A. SECONDARY DATA (PREDICTOR VARIABLES)

Demographic Indicators:

Total population (1988)
Rate of growth of population (1988)
Total agricultural population (1988)
Economically active population in agriculture (1988)
Estimated percent women in agriculture (1988)
Percentage of labour force in agriculture (1988)
Life expectancy at birth (1988)

Politico-Economic Indicators:

Independence date
GDP per capita 1988 (U.S. Dollars)
Agricultural gross domestic product (1988)

Agricultural Production Indicators:

Total cereal yield in kg/ha (1988)

Communication Indicators:

Adult literacy percentage (1988)
Number of Radio / 1000 inhabitants (1988)
Number of Television / 1000 inhabitants (1988)
Number of News Paper / 1000 inhabitants (1988)

Agricultural Input Indicators:

Arable land (1988)
Cropped land irrigated (1988)
Cropped land per capita (1988)

Land under permanent crops (1988)
Total fertilizer consumption in kg/ha (1988)
Tractor density (1988)

Food Consumption Indicators:

Cereal consumption per capita
Daily calorie supply per capita (1988)
Daily grams of protein supply per capita (1988)

QUESTIONNAIRE PART 2

B. PRIMARY DATA OF HUMAN RESOURCES

Number of extension staff over time:

Section B1: In this section we would like to determine the total number of professional and technical staff in extension, including all field officers and assistants, since 1960. For this table, do not include clerical staff, secretaries, security personnel, drivers and so forth. Only include those professional and technical professional and technical personnel that help plan, implement and evaluate extension programmes.

Table B-1: Total Number of Professional and Technical Extension Personnel

Year	*Total Number of Extension Personnel*	*Year*	*Total Number of Extension Personnel*
1960		1983	
1965		1984	
1970		1985	
1975		1986	
1980		1987	
1981		1988	
1982			

Section B2: *Educational Qualification of Extension Personnel over Time*

Given the total number of Professional and technical extension personnel listed in Table B-1, in this section please indicate the number of extension personnel who hold the following minimum educational qualifications:

1. Less than secondary or technical secondary school qualification.
2. A secondary or technical secondary school qualification or its equivalent.
3. A post-secondary, intermediate-level qualification (*i.e.* certificate or diploma).
4. A university degree (*e.g.* B.Sc. degree) or its equivalent.
5. A post-graduate (research) qualification (*i.e.* a M. Sc. or Ph. D. degree) or its equivalent. Please provide these data on the educational qualifications of your extension staff for as many years as they are available.

Table B-2: Educational Qualification of Professional and Technical Extension Personnel

Year	*Less than Secondary School*	*Secondary School or Equivalent*	*Intermediate Level (Certi. or Diploma)*	*University Graduate (B. Sc. or Equivalent)*	*Post-Graduate Degree (M.Sc. or Ph. D.)*	*Total Ext. Staff*
1960						
1965						
1970						
1975						
1980						
1985						
1986						
1987						
1988						

Section B3: *Extension Personnel by Programme Category and Gender*

In this section we would like to determine the current (1988) distribution of extension personnel by function and/or programme category, as well as by gender. First, please indicate the total number of extension officers and assistants that would be classified in each of the major functional categories of (1) administrators and supervisors, (2) technical or subject-matter specialists (SMSs), and (3) field officers and assistants. Then differentiate these personnel by programme area and then by gender. Please do not include clerical, secretarial, security, drivers and other non-technical support personnel in this table. The total number of non-technical support personnel should be included at the bottom of this table, in the space provided. When you have completed this table, be sure the columns and rows total correctly and that the annual total corresponds with those totals found in table 1 and 2 for 1988. If for any reason 1988 data are not available, please give the same information for the most recent years when data are available and specify which year.

Table B-3: Current Number of Extension Personnel Distributed by Function, Programme Area and Gender

Extension Personnel by Function, Programme and Gender		*Male*	*Female*	*Total*
A	Administrators/Supervisors			
B	Extension Specialists			
B1	Technical/Subject Matter (SMS)			
B2	Agric. Information Specialists*			
B3	Training Specialist**			
	Sub-Total Specialists			
C	Field Extension Officers/ Assistants			
C1	Agricultural Extensionists			
C2	Home Economics Extensionists			
C3	Rural Youth Extensionists			
C4	Rural Development Extensionists			
C5	Other (specify)			
	Sub-Total Field Extension Staff			
	Total Number of Extension Staff (A+B+C)			
D	Other Extension Support Personnel (*e.g.* Secretaries, Drivers, Clerks)			

* Staff at the Agricultural Information Section (Central Level).

** Staff at the Central and Regional Agricultural Training Centre.

*** Staff at the Central level office.

Some extension worker is responsible for implementing agricultural extension, rural youth, training and Agricultural Information Programme at the district level.

Table B-4: Information about other Extension Personnel

1. Approximately how many professional and technical extension personnel left the extension organization during the most recent calendar or fiscal year due to retirements, deaths, resignations or any other reason?
   Reporting Period: 19... Departing Personnel: .... Persons
2. How many new professional and technical level extension staff were hired during this same period?
   New Personnel: .....Persons
3. Approximately how many professional and technical extension positions remained vacant at the end of this same reporting period?
   ....vacancies
4. What was the primary reason for these vacancies not being filled? (*i.e.* lack of trained people, lack of budget, etc.)
   ____
   ____

Table B-5: Current Educational Qualifications and Average Monthly Salary of Extension Personnel by Category of Position

Please indicate how many years of formal education (on the average, including primary school), and what degree, diploma or certificates most extension personnel are now expected to have when *entering* the agricultural extension organization; or when being *promoted* into each of the following types of positions. Also indicate the current average *monthly* salary level for each type of position.

Type of Position	*Average years of formal education**	*Type of diploma, degree or certificate generally needed for this type of position*	*Average monthly salary level in local currency†*
Administrative or Supervisory Officer	______Years	______	______
Technical or Subject Matter Specialists	______Years	______	______
Extension Officer/Agent or Instructor	______Years	______	______
Village level worker or extension assistants	______Years	______	______

* Please include total number of years of primary, secondary and tertiary education.
† Please give salary in the local currency unit.
What is local currency ____________

QUESTIONNAIRE PART 3

Primary Data of Financial Resources

Section C1: *Financial Resources Expended by Extension*

In this section we would like to determine the annual level of recurrent extension expenditures for selected years. Here we are primarily interested in regular extension expenditures for all major recurring categories of expenditures, such as salaries and personal emoluments, programme and operating costs, and routine capital equipment replacement and maintenance. Do not include in this table major, one time only, capital expenditure such as a new building, as this information will be requested later.

To put this financial information into perspective, we would also like to know the overall size of the annual, recurrent Ministry of Agriculture budget for these same years. With this information, we can determine what proportion of the ministry's budget is going to extension and whether extension's share has remained stable, increased or decreased over time. Please report these financial data in your national currency unit, for the dates indicated, for as far back as you have accurate records. If there are years when these data are unavailable, please leave these spaces blank. However these data therefore, be as complete as possible.

Table C-1: Annual Recurrent Extension and Ministry Expenditures

Please give total annual Extension and Ministry expenditures in millions of the national currency unit.

The national currency unit is:_____________

For Fiscal Year	*Total Annual Extension Expenditures (recurrent costs in 000,000)*	*Total Annual Ministry of Agriculture expenditures (recurrent costs in 000,000)*
1960		
1965		
1970		
1975		
1980		
1985		
1986		
1987		
1988		

Section C2: *Annual Extension Expenditures by Cost Element*

In this table, please break-down the annual, recurrent extension expenditure data into three major cost element or categories, which are defined briefly below. These categories are (1) salaries costs, including all personnel emoluments, (2) programme and other operating costs, and (3) recurring capital expenditures (*Note*— do not include large, one time only, capital expenditures as they are required in Table C2 below). Please provide these cost breakdowns for as many years as possible, especially the most recent years.

Table C2: Extension Expenditures by Cost Element (Salaries, Programmes and Capital; see definitions below)

(Please give expenditures in millions of the national currency unit)

For Fiscal Year	*Salary Costs (000,000)*	*Programme Costs (000,000)*	*Capital Costs (000,000)*	*Other or Unlocated Costs (000,000)*	*Total Recurrent Extension Costs (000,000)*
1960					
1965					
1970					
1975					
1980					
1985					
1986					
1987					
1988					

Definitions

"Salary costs" includes all salary and wage payments, plus all fringe benefits and personnel emoluments, paid directly to or on behalf of all. extension personnel.

"Programme costs" include all regular programme expenses and operating costs, such as transportation, fuel, perdiem and other travel expenses, office supplies, printing and photographic expenses, supplies for field demonstrations, field days and fair exhibits, communications expenses, including telephone and postage, and so forth.

"Capital costs" refer to the purchase or replacement (and maintenance) of equipment, vehicles and other capital items, as well as the maintenance of buildings and grounds that are funded out of the regular, annual recurrent budget for the extension organization.

Section C3: *Major Capital Expenditures*

Many developing countries have received bilateral and multilateral donor support to strengthen the human physical resources of the national extension system. These major capital investment generally involve multi-year projects, concentrating on such areas as training, equipment, buildings and technical assistance. If possible, since 1975, could you list all of the major capital improvement projects funded by your government and/or by an external donor to strengthen the human and physical infrastructure of the extension organization. If more space is needed, please attach additional page, using the same format.

Table C3 : Major Capital Investment in Extension Since 1975
(Please specify which currency is being used for each project)

Project Name I	*Example Project Agri. Ext. Research Programme I*	*Project 1 Hill Food Production Programme*	*Project 2 Hill Agri. Dev. Project*	*Project 3 Agri. Ext. Project II*
Sources of funds (*i.e.* Govt., or specify which Donor?)	IDA/UNDP	IDA/UNDP	ADB/M	IDA/UNDP
In which currency is budget given?				
Year Project started?				
Year of Completion				

Expenditure by category
(Please specify if in thousands or in millions)

Funding Categories:

1.	Training				
2.	Equipment / commodities (except vehicles)				
3.	Vehicles				
4.	Building Construction				
5.	Technical Assistance				
6.	Other (specify)				
Total Capital Expenditures					

Section C4: *Ministry of Agriculture Expenditures by Functional Units or Areas*

In this section, we would like to determine how the Ministry of Agriculture's budget is allocated to different departments or functional units that serve the agricultural sector in your country. We have listed some major functional areas that may receive funds, but since Ministry is organized differently, we have left some categories open so you can fill in the name and budget for all of the major functional units of the Ministry. Please give these data for the most recent fiscal year and in millions of your national currency unit.

Table C-4: Breakdown of Ministry of Agriculture Expenditures for Fiscal Year: 19…

Functional Department or Unit	*Expenditure in (000,000) of the national currency unit*
General Ministry Operations and Overhead	
Agricultural Extension	
Agricultural Research	
Intermediate and Higher Agricultural Education and Training	
Agricultural Planning, including Agricultural census, statistics, crop forecasting etc.	
Commodity price and income support programme's	
Input supply and subsidiary programmes to reduce cost of production inputs for farmers	
Others (please specify)	
Total Ministry expenditure	

APPENDIX B

LIST OF PARTICIPATING COUNTRIES BY REGION

Africa
Algeria
Benin
Burkina Faso
Burundi
Cameroon
Cape Verde
Congo
Cote d'Ivoire
Ethiopia
Gambia
Ghana
Guinea-Bissau
Guinea Rep. of
Kenya
Malawi
Mali
Mauritania
Mauritius
Morocco
Mozambique
Niger
Rwanda
Senegal
Sierra Leone
Sudan
Tanzania
Tunisia
Uganda
Zaire
Zambia
Zimbabwe

Asia and the Pacific
Australia
Bangladesh
Bhutan
China
Cook Island
Fiji
India
Indonesia
Japan
Korea, Republic of
Laos
Malaysia
Mongolia
Myanmar, Union of
Nepal
New Zealand
Pakistan
Papua New Guinea
Philippines
Solomon Islands
Sri Lanka
Thailand
Tonga
Vanuatu
Vietnam
Western Samoa

Europe
Belgium
Czechoslovakia
Denmark
Finland
Germany, Fed. Rep. of
Hungary
Iceland
Ireland
Malta
Netherlands
Norway
Portugal
Sweden
United Kingdom

Latin America and the Caribbean
Argentina
Bolivia
Brazil
Chile
Colombia
Costa Rica
Dominica
Dominican Republic
Ecuador
El Salvador
Guatemala
Guyana
Haiti
Honduras
Jamaica
Mexico
Nicaragua
Panama
Peru
Saint Kitts and Nevis
Saint Lucia
Suriname
Trinidad and Tobago
Uruguay

North America
Bermuda
Canada
United States of America

The Near East
Afghanistan
Bahrain
Cyprus
Egypt
Iraq
Israel
Jordan
Lebanon
Oman
Qatar
Saudi Arabia
Somalia
Syria
Turkey
United Arab Emirates
Yemen, PDR

APPENDIX C

DATA OF DIFFERENT VARIABLES

Table C-1: Total Cereal Yield* (kg/ha) by Geographic Region and Economic Group

Geographic Region	*Economic Group (PCI)*				*Region*	*No. of Countries*
	Low Income	*Lower Middle Income*	*Upper Middle Income*	*Higher Income*		
Africa	1,042 (21)	1,233 (9)	671 (1)	—	1,085	(31)
Asia	2,396 (11)	2,288 (8)	5,844 (1)	3,967 (3)	2,713	(23)
Europe	—	—	3,361 (2)	4,841 (11)	4,613	(13)
Latin America	1,629 (2)	2,193 (15)	2,242 (6)	—	2,209	(23)
North America	—	—	—	1,390 (2)	1,390	(2)
The Near East	1,244 (3)	2,086 (6)	—	2,799 (6)	2,203	(15)
Group Total	1,493 (37)	1,969 (38)	2,789 (10)	3, 851 (22)	2,268	(107)

Note—The number of countries are given in the parentheses.

* Total Cereal Yield (kg/ha) was taken from the FAO Production Year Book, 1989 for the year 1988 of all responding countries. The data of seven countries were not available.

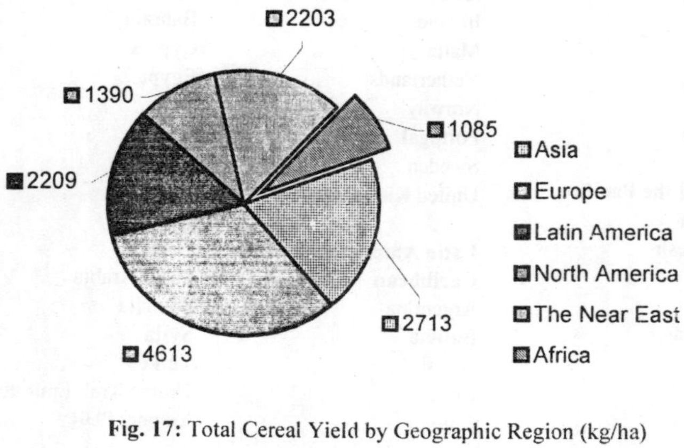

Fig. 17: Total Cereal Yield by Geographic Region (kg/ha)

Table C-2: Farmer Income* per Farmer by Geographic Region and Economic Group

Geographic Region	*Economic Group (PCI)*				*Region*	*No. of Countries*
	Low Income	*Lower Middle Income*	*Upper Middle Income*	*Higher Income*		
Africa	145 (18)	631 (8)	—	—	295	(26)
Asia	198 (7)	1,161 (5)	2,295 (1)	15,939 (3)	3,582	(16)
Europe	—	—	2,826 (2)	17,361 (8)	14,454	(10)
Latin America	257 (2)	670 (12)	1,468 (6)	—	868	(20)
North America	—	—	—	32,234 (2)	32,234	(2)
The Near East	—	2,471 (3)	—	8,720 (3)	5,730	(6)
Group Total	167 (27)	969 (28)	1,862 (9)	17,333 (16)	4,071	(80)

Note — The number of countries are given in the parentheses.
* Farmer income was computed by using the agricultural gross domestic product for the year 1988 deflated and converted into 1980 U.S. Dollar. This Dollar amount was divided by the economically active population in Agriculture to provide a measure of average farmer income for each country.

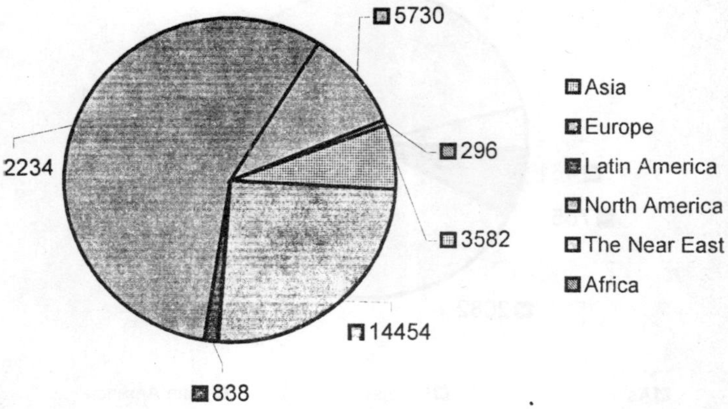

Fig. 18: Farmer Income by Geographic Region (US $)

Table C-3: Agricultural Income* per Farmer by Geographic Region and Economic Group

Geographic Region	*Economic Group (PCI)*				*Region*	*No. of Countries*
	Low Income	*Lower Middle Income*	*Upper Middle Income*	*Higher Income*		
Africa	237 (18)	510 (8)	—	—	321	(26)
Asia	466 (7)	11,900 (5)	5,585 (1)	7,170 (3)	2,941	(16)
Europe	—	—	504 (2)	2,476 (8)	2,082	(10)
Latin America	330 (2)	896 (12)	647 (6)	—	765	(20)
North America	—	—	—	451 (2)	451	(2)
The Near East	—	1,062 (3)	—	7,021 (3)	4,041	(6)
Group Total	303 (27)	983 (28)	1,164 (9)	3,955 (16)	1,368	(80)

Note—The number of countries are given in the parentheses.

* Agricultural income /ha of arable land was computed by using the agricultural gross domestic product for the year 1988 deflated and converted into 1980 U.S. Dollar. This Dollar amount was divided by the number of hectares of arable land in each country, to provide a general measure of agricultural income/ha for each country.

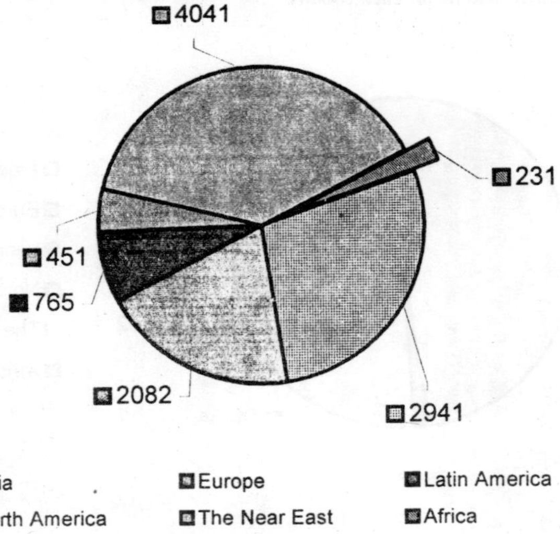

Fig. 19: Agricultural Income by Geographic Region (US $ per Hectare Arable Land)

Table C-4: Total Fertilizer Consumption *(Nitrogen, Phosphorus and Potash kg/ha Arable and Permanent Cropland) by Geographic Region and Economic Group

Geographic Region	*Economic Group (PCI)*				*Region*	*No. of Countries*
	Low Income	*Lower Middle Income*	*Upper Middle Income*	*Higher Income*		
Africa	9 (19)	50 (8)	23 (1)	—	21	(28)
Asia	76 (11)	68 (6)	400 (1)	393 (3)	134	(21)
Europe	—	—	177 (2)	543 (12)	491	(14)
Latin America	16 (2)	81 (15)	72 (7)	—	73	(24)
North America	—	—	—	70 (2)	70	(2)
The Near East	3 (3)	117 (6)	-	213 (7)	138	(16)
Group Total	30 (35)	78 (35)	116 (11)	389 (24)	137	(105)

Note—The number of countries are given in the parentheses.

* Total Fertilizer consumption (Nitrogen, Phosphorus, and Potash kg/ha arable land and permanent crops) was taken from the FAO Fertilizer Year Book 1989 for the year 1988 of all responding countries.

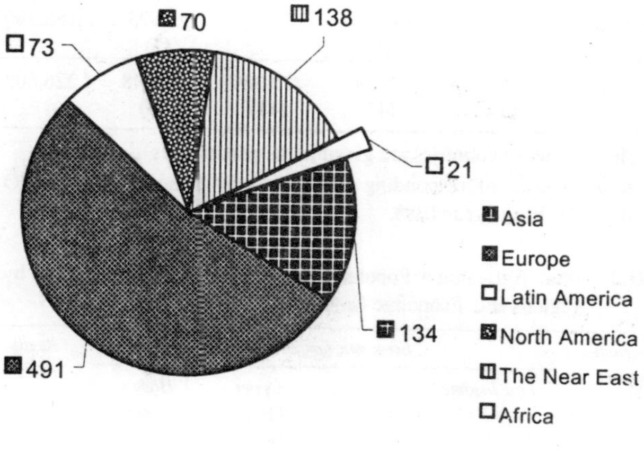

Fig. 20: Total Fertilizer Consumption by Region (NPK kg/ha Arable and Permanent Cropland)

APPENDIX D

BASIC DATA TABLES

Table D-1: Total Population of the Responding Countries by Geographic Regions and Economic Group (000).

Geographic Region	*Economic Group (PCI)*				*Region*	*No. of Countries*
	Low Income	*Lower Middle Income*	*Upper Middle Income*	*Higher Income*		
Africa	267,373 (21)	73,438 (9)	23,841 (1)	—	364,652	(31)
Asia	2,464,973 (11)	155,214 (11)	42,262 (1)	142,313 (3)	2,804,946	(26)
Europe	—	—	20,843 (2)	186,042 (12)	206,885	(14)
Latin America	7,270 (2)	201,126 (15)	183,059 (7)	—	391,445	(24)
North America	—	—	—	272,068 (3)	272,068	(3)
The Near East	24,334 (3)	140,337 (6)	—	21,925 (7)	2,186,596	(16)
Group Total	2,763,770 (37)	570,115 (41)	270,369 (11)	662,348 (25)	4,226,602	(114)

Note—The number of countries are given in the parentheses.

* Total population of responding countries was taken from the FAO Production Yearbook, 1988 for the year 1988.

Table D-2: Total Agricultural Population of the Responding Countries by Geographic Regions and Economic Groups (000)

Geographic Region	*Economic Group (PCI)*				*Region*	*No. of Countries*
	Low Income	*Lower Middle Income*	*Upper Middle Income*	*Higher Income*		
Africa	199,141 (21)	37,938 (9)	5,989 (1)	—	243,068	(31)
Asia	1,586,623 (11)	68,473 (7)	10,376 (1)	9,606 (3)	1,675,078	(22)
Europe	—	—	3,284 (2)	7,514 (12)	10,798	(14)
Latin America	4,069 (2)	66,191 (15)	41,627 (7)	—	111,887	(24)
North America	—	—	—	7,945 (3)	7,945	(3)
The Near East	14,250 (3)	53,429 (6)	—	6,341 (6)	74,020	(15)
Group Total	1,804,083 (37)	226,031 (37)	61,276 (11)	31,406 (24)	2,122,796	(109)

Note — The number of countries are given in the parentheses.

* Total agricultural population of responding countries was taken from the FAO Production Yearbook, 1988 for the year 1988.

Table D-3: Number of Farmers* of the Responding Countries by Geographic Regions and Economic Groups (000)

Geographic Region	*Economic Group (PCI)*				*Region*	*No. of Countries*
	Low Income	*Lower Middle Income*	*Upper Middle Income*	*Higher Income*		
Africa	84,975 (21)	14,147 (9)	1,379 (1)	—	100,501	(31)
Asia	772,893 (11)	31,755 (11)	4,859 (1)	4,980 (3)	814,487	(25)
Europe	—	—	1,473 (2)	5,984 (12)	7,457	(14)
Latin America	1,913 (2)	21,516 (14)	15,196 (7)	—	38,625	(23)
North America	—	—	—	3,531 (3)	3,531	(3)
The Near East	4,736 (3)	19,379 (6)	—	1,900 (7)	26,015	(16)
Group Total	864,517 (37)	86,797 (39)	22,907 (11)	16,395 (25)	990,616	(112)

Note—The number of countries are given in the parentheses.

* The economically active population in agriculture of all responding countries was assumed as number of farmers. The data was taken from the FAO Production Yearbook, 1988 for the year 1988.

Table D-4: Arable Land of Responding Countries by Geographic Regions and Economic Groups (000)

Geographic Region	*Economic Group (PCI)*				*Region*	*No. of Countries*
	Low Income	*Lower Middle Income*	*Upper Middle Income*	*Higher Income*		
Africa	71,164 (21)	28,133 (9)	6,970 (1)	—	106,267	(31)
Asia	324,079 (11)	25,173 (11)	1,997 (1)	51,463 (3)	402,712	(26)
Europe	—	—	7,102 (2)	30,655 (12)	37,757	(14)
Latin America	1,035 (2)	45,755 (15)	94,359 (7)	—	141,149	(24)
North America	—	—	—	233,786 (3)	233,786	(3)
The Near East	9,043 (3)	31,023 (6)	—	1,608 (7)	41,674	(16)
Group Total	405,321 (37)	130,084 (41)	110,428 (11)	110,428 (25)	963,345	(114)

Note— The number of countries are given in the parentheses.

* Arable land of responding countries was taken from the FAO Production Yearbook, 1989 for the year 1988.

Table D-5: Percentage of Low Level Educated Extension Personnel* in Responding Countries by Geographic Region and Economic Group

Geographic Region	*Economic Group (PCI)*				*Region*	*No. of Countries*
	Low Income	*Lower Middle Income*	*Upper Middle Income*	*Higher Income*		
Africa	73.3 (21)	76.4 (9)	76.9 (1)	–	74.3	(31)
Asia	77.8 (11)	73.4 (11)	69.0 (1)	35.1 (3)	70.7	(26)
Europe	–	–	45.0 (2)	16.7 (12)	20.7	(14)
Latin America	87.5 (2)	58.3 (15)	56.1 (7)	–	60.1	(24)
North America	–	–	–	9.7 (3)	9.7	(3)
The Near East	66.4 (3)	44.6 (6)	–	39.9 (7)	46.6	(16)
Group Total	74.8 (37)	64.3 (41)	57.1 (11)	24.6 (25)	58.3	(114)

Note—The number of countries are given in the parentheses.

* Less than University level education of extension personnel is categorized as low level education.

APPENDIX E

SUMMARY OF RESULTS IN TWO SETS OF ANALYSES

Table E-1: Significance of Quantity of Human Resources with Agricultural Development Indicators for all Responding Countries ($N = 114$)

	Variables	*N*	χ^2	*ANOVA of Log P*
1.	Farmers/Researcher *vs.* Cereal Yield	95	.00000	.0000
2.	Farmers/Researcher *vs.* Farmer Income	84	.00000	.0000
3.	Farmers/Researcher *vs.* Agricultural Income	84	.01294	.1874
4.	Farmers/Researcher *vs.* Fertilizer Consumption	93	.00000	.0000
5.	Farmers/EW *vs.* Cereal Yield	99	.00167	.0000
6.	Farmers/EW *vs.* Farmer Income	85	.00001	.0000
7.	Farmers/EW *vs.* Agril Income	85	.00453	.0036
8.	Farmers/EW *vs.* Fertilizer Consumption	98	.00013	.0000
9.	Arable Land/R *vs.* Cereal Yield	97	.00000	.0000
10.	Arable Land/R *vs.* Farmer Income	85	.00000	.0000
11.	Arable Land/R *vs.* Agril Income	85	.00000	.0000
12.	Arable Land/R *vs.* Fertilizer Consumption	94	.00000	.0004
13.	Arable Land/ EW *vs.* Cereal Yield	101	.15089	.2892
14.	Arable Land/ EW *vs.* Farmer Income	86	.19943	.0847
15.	Arable Land/ EW *vs.* Agril Income	86	.00010	.0003
16.	Arable Land/ EW *vs.* Fertilizer Consumption	99	.03008	.7686
17.	Percentage of SMS *vs.* Cereal Yield	95	.31888	.8861
18.	Percentage of SMS *vs.* Farmer Income	83	.21912	.2701
19.	Percentage of SMS *vs.* Agril Income	83	.12163	.2863
20.	Percentage of SMS *vs.* Fertilizer Consumption	93	.62492	.0685
21.	Percentage of FEW *vs.* Cereal Yield	82	.10229	.0428
22.	Percentage of FEW *vs.* Farmer Income	72	.24352	.0741
23.	Percentage of FEW *vs.* Agril Income	72	.27225	.1539
24.	Percentage of FEW *vs.* Fertilizer Consumption	80	.11426	.0129
25.	EW/R *vs.* Cereal Yield	88	.03805	.0218
26.	EW/R *vs.* Farmer Income	79	.01080	.0967
27.	EW/R *vs.* Agril Income	79	.16421	.4887
28.	EW/R *vs.* Fertilizer Consumption	85	.00961	.0006

Table E-2: Significance of Quantity of Human Resources with Agricultural Development Indicators only for Developing Countries ($N = 89$)

	Variables	*N*	χ^2	*ANOVA of Log P*
1.	Farmers/Researcher *vs.* Cereal Yield	77	.00226	.0012
2.	Farmers/Researcher *vs.* Farmer Income	61	.00010	.0758
3.	Farmers/Researcher *vs.* Agril. Income	61	.11970	.6798
4.	Farmers/Researcher *vs.* Fertilizer Consumption	74	.00085	.0003
5.	Farmers/EW *vs.* Cereal Yield	80	.00847	.0011
6.	Farmers/EW *vs.* Farmer Income	62	.00652	.0136
7.	Farmers/EW *vs.* Agril. Income	62	.52125	.2361
8.	Farmers/EW *vs.* Fertilizer Consumption	77	.00070	.0000
9.	Arable Land/R *vs.* Cereal Yield	79	.00448	.0004
10	Arable Land/R *vs.* Farmer Income	62	.00874	.0017
11.	Arable Land/R *vs.* Agril Income	62	.00009	.0001
12.	Arable Land/R *vs.* Fertilizer Consumption	75	.00001	.0040
13.	Arable Land/ EW *vs.* Cereal Yield	82	.10167	.0029
14.	Arable Land/ EW *vs.* Farmer Income	63	.41617	.0824
15.	Arable Land/ EW *vs.* Agril Income	63	.00014	.0001
16.	Arable Land/ EW *vs.* Fertilizer Consumption	78	.01708	.6199
17.	Percentage of SMS *vs.* Cereal Yield	79	.10108	.4677
18.	Percentage of SMS *vs.* Farmer Income	61	.95044	.3399
19.	Percentage of SMS *vs.* Agril Income	61	.18162	.1709
20.	Percentage of SMS *vs.* Fertilizer Consumption	75	.76425	.4628
21.	Percentage of FEW *vs.* Cereal Yield	70	.81460	.4767
22	Percentage of FEW *vs.* Farmer Income	56	.59590	.3903
23.	Percentage of FEW *vs.* Agril Income	56	.61758	.4887
24.	Percentage of FEW *vs.* Fertilizer Consumption	67	.27032	.0158
25.	EW/R *vs.* Cereal Yield	75	.33014	.2394
26.	EW/R *vs.* Farmer Income	61	.00023	.3605
27.	EW/R *vs.* Agril Income	61	.13090	.4539
28.	EW/R *vs.* Fertilizer Consumption	71	.00744	.0003

Table E-3: Significance of Quality of Human Resources with Agricultural Development Indicators for all Responding Countries ($N = 114$)

	Variables	*N*	χ^2	*ANOVA of Log P*
1.	Percentage of EW with University Degree or Higher Education *vs.* Cereal Yield	95	.02806	.0005
2.	Percentage of EW with University Degree or Higher Education *vs.* Farmer Income	79	.00919	.1194
3.	Percentage of EW with University Degree or Higher Education *vs.* Agril Income	79	.41051	.6711
4.	Percentage of EW with University Degree or Higher Education *vs.* Fertilizer Consumption	92	.01733	.0006

Table E-4: Significance of Quality of Human Resources with Agricultural Development Indicators of only Developing Countries ($N = 89$)

	Variables	*N*	χ^2	*ANOVA of Log P*
1.	Percentage of EW with University Degree or Higher Education *vs.* Cereal Yield	79	.02574	.0515
2.	Percentage of EW with University Degree or Higher Education *vs.* Farmer Income	60	.26787	.4230
3.	Percentage of EW with University Degree or Higher Education *vs.* Agril Income	60	.44645	.1352
4.	Percentage of EW with University Degree or Higher Education *vs.* Fertilizer Consumption	75	.70950	.0407

Table E-5: Significance of Financial Resources with Agricultural Development Indicators for all Responding Countries ($N = 114$)

	Variables	*N*	χ^2	*ANOVA of Log P*
1.	Extension Expen./F *vs.* Cereal Yield	80	.00004	.0533
2.	Extension Expen./F *vs.* Farmer Income	68	.00000	.3139
3.	Extension Expen./F *vs.* Agril Income	68	.00010	.4018
4.	Extension Expen./F *vs.* Fertilizer Consumption	80	.00001	.2808
5.	Research Expen./F *vs.* Cereal Yield	84	.00000	.8670
6.	Research Expen./F *vs.* Farmer Income	76	.00000	.4787
7.	Research Expen./F *vs.* Agril Income	76	.00000	.5219
8.	Research Expen./F *vs.* Fertilizer Consumption	84	.00000	.1712

Table E-6: Significance of Financial Resources with Agricultural Development Indicators of only Developing Countries ($N = 89$)

Variables		*N*	χ^2	*ANOVA· of Log P*
1.	Extension Expen./F *vs.* Cereal Yield	66	.39161	.8670
2.	Extension Expen./F *vs.* Farmer Income	53	.00000	.4787
3.	Extension Expen./F *vs.* Agril Income	53	.36463	.5219
4.	Extension Expen./F *vs.* Fertilizer Consumption	64	.00950	.1712
5.	Research Expen./F *vs.* Cereal Yield	66	.19205	.0533
6.	Research Expen./F *vs.* Farmer Income	54	.00050	.3139
7.	Research Expen./F *vs.* Agril Income	54	.26253	.4018
8.	Research Expen./F *vs.* Fertilizer Consumption	65	.02017	.2808

Table E-7: Summary of Multiple Regression

Human Resources

Variables		*All Countries*		*Developing Countries*	
		Beta	*p*	*Beta*	*p*
Cereal Yield:					
a)	FARMER/R	–.0560	.6396	–.1305	.3739
b)	ARABLE/EW	–.1829	.0748	–.1981	.1822
c)	SMS %	.2127	.0570	.1937	.1463
d)	HIGHED %	.3298	.0043	.1928	.1524
e)	FARMER/EW	–.2412	.0453	–.1603	.3642
Farmer Income:					
a)	FARMER/R	–.0495	.6551	–.1213	.3936
b)	SMS %	.3214	.0046	.0470	.7600
c)	ARABLE/EW	.2431	.0180	–.3747	.0298
d)	FARMER/EW	–.2261	.0378	-1.969	.9991
e)	HIGHED %	.2526	.0296	.4465	.0061
Agricultural Income:					
a)	FARMER/R	–.0483	.7327	–.2281	.1325
b)	SMS %	–.0634	.6517	.3335	.0450
c)	ARABLE/EW	–.0246	.8479	–.1584	.3755
d)	FARMER/EW	–.0129	.9243	–.1373	.4087
e)	HIGHED %	–.0264	.8580	–.1472	.4411
Fertilizer Consumption:					
a)	FARMER/R	.0125	.9177	–.1017	.4988
b)	ARABLE/EW	–.2119	.0417	–.2334	.1238
c)	SMS %	.3744	.0012	.2803	.0429
d)	HIGHED %	.2364	.0407	.0345	.8003
e)	FARMER/EW	–.2484	.0422	–.1818	.3185

Table E-8: Summary of Multiple Regression

Financial Resources

Variables	*All Countries*		*Developing Countries*	
	Beta	*p*	*Beta*	*p*
Yield Cereal:				
Total R&E Expenditure/Farmer	.5635	.0000	.4267	.0026
Farmer Income:				
Total R&E Expenditure/Farmer	–.6929	.0000	.4267	.0035
Agricultural Income:				
Total R&E Expenditure/Farmer	.4515	.0008	.4723	.0011
Fertilizer Consumption:				
Total R&E Expenditure/Farmer	.6688	.0000	.4293	.0015

BIBLIOGRAPHY

Agency for International Development. *Niger Cereals Research Project No. 68-0225.* Unclassified project paper, Washington, DC, October 1981.

Ames, G. C. W. (1988). Human capital, agricultural development and the African food crisis. *Agricultural Administration and Extension, 28*(1) pp. 1-17.

Axinn, G. H., & Thorat, S. (1972). *Modernizing world agriculture — A comparative study of agricultural extension education systems.* NY: Praeger Publishers.

Azrag, E.S.E. (1987). *Future Trends in Agricultural Production within the Arab World* [Computer Programme]. CAB: OW weed abstracts 1988 037-01465. AN: G087020.

Belloncle, G. (1989). Proposal for a new approach to extension services in Africa. In N. Roberts (Ed.), *Agricultural Extension in Africa* collectively—*A World Bank Symposium* (pp. 37-44). Washington DC: The World Bank.

Bembridge, T. J. (1989). An evaluation of the Venda Agricultural Extension Service. *South African Journal of Agricultural Extension*, pp. 19-27.

Benor, D. (1987). Training and Visit Extension: Back to basics. In William M. Rivera & Susan G. Schram (Eds.), *Agricultural Extension Worldwide Issues, Practices and Emerging Priorities* (pp. 137-148). London: Croom Helm.

Benor, D., Harrison, J. Q., and Baxter, M. (1984). *Agricultural extension: The training and visit system.* Washington, DC: World Bank.

Berger, M., Delancey, V., & Mellencamp, A. (1984). *Bridging the Gender gap in Agricultural Extension* [Computer Programme]. CAB: or World Agricultural Economics & Rural Sociology Abstracts, AN: R195408.

Bouldiung, E. (1977). *Women in the twentieth century world.* NY: Sage Publications.

Bramblett, L. R. & Flewellen, W. C., Jr. (1976). Evaluation and reward systems for continuing education and extension professionals: A national review. *The NUEA Spectator, 36-40*, 9-12.

Brisset, C. (1986). *Irrigation: The absolute Imperative* [Computer Programme]. CAB: OR World Agricultural Economics & Rural Sociology Abstracts 1987. AN: R 436871.

Brown, W. M., Jr. (1983). Constraints in effective extension programme development in Third World countries. In J. B. Claar & L. H. Watts (Eds.), *Knowledge transfer in developing countries: Status, constraints, outlook* (pp. 42-44). Urbana: University of Illinois at Urbana-Champaign, International Program for Agriculture Knowledge Systems.

Central Intelligence Agency (1989). *The World Factbook 1989.* Central Intelligence Agency. Washington, DC.

Chaudhary, M. A. & Al-Haj, F. M. (1985). Agricultural education and extension in developing countries. *Agricultural Administration, 20*(3), pp. 169-186.

Clark, Richard W. (1987). Human Resource Development: Key to Extension's Survival. *Journal of Extension, XXV*, 25-26.

Contado, T. E. (1990). What FAO's Studies Reveal. In B. E. Swanson (Ed.). *Global Consultation on Agricultural Extension—Report.* Food and Agricultural Organization of the United Nations, Rome, Italy.

Deutsch, K. W. (1980). On the utility of indicator systems, in C. L. Taylor (Ed.). *Indicators systems for political, economic and social analysis*, (pp. 11-21). Cambridge, MA: Oelgeschlager, Gunn, & Hain.

Doorenbos, J., Haverkort, B., & Jiggins, J. (1988). Women and Rationalization of Smallholder Agriculture. *Agricultural Administration and Extension*, 28(2), 101-112.

Dutia, B. P. (1990). The Challenges ahead for Agricultural Extension (Key note address). In B. E. Swanson (Ed.), *Report of the Global Consultation on Agricultural Extension*, (pp. 34-42). FAO, Rome.

Europa publications (1990). *The Europa World Year Book 1990.* London: Author.

Esman, Milton Jacob (1980). *Paraprofessional in Rural Development.* NY: Ithaca.

Esman, Milton J. (1983). *Paraprofessionals in Rural Development—Issues in Field-level Staffing for Agricultural Projects* (World Bank Staff Working Papers Number 573). Washington, DC: The World Bank.

Evenson, Robert E. (1987). The IARCs and their impact on national research and extension programs. In William M. Rivera & Susan G. Schram (Eds.), *Agricultural Extension Worldwide—Issues, Practices and Emerging Priorities* (pp. 115-136). London: Croom Helm.

Evenson, Robert E. (1988, p. 22), "The economics of extension". *Bulletin 24 Reading Rural Development Communications*, University of Reading, Agricultural Extension and Rural Development Centre London Road Reading England.

FAO (1984). Economic and Social Development Paper 40. *Socio-economic indicators relating to the agricultural sector and rural development.* Rome.

FAO (1985). *Women in agricultural production; Women in food production and food security in Africa, and Women in developing agriculture, Women in Agriculture Series Nos. 1, 3, 4*, Rome.

FAO (1986). *Worldwide Estimates and projections of the agricultural and non-agricultural population segments*, 1950-2025. Rome: Author.

FAO (1988). *FAO Production Year Book 1988.* FAO Statistics No. 88. Vol. 42, Rome, Author.

FAO (1989). *FAO Production Year Book 1989.* FAO Statistics No. 94. Vol. 43, Rome, Author.

FAO (1989). *FAO Fertilizer Year Book 1989.* FAO Statistics No. 95. Vol. 39, Rome, Author.

Fortmann, L. (1985). Factors Affecting Agricultural and other Rural Extension Services in Botswana. *Agricultural Administration, 18,* 13-23.

Gabriel, T. (1989). *Women and Extension Services* [Computer Programme]. CAB Abstracts 1987-89, AN. R739360.

Gintings, M. R. (1982). *A Study of Empathic Ability of Field Extension Workers in Java, Indonesia* [Computer Programme]. CAB: 1R Rural Extension, Education and Training Abstracts 1984 007-00817.

Hayward, J. A. (1990). Agricultural Extension: The World Bank's Experience and Approaches. In B. E. Swanson (Ed.). *Report of the Global Consultation on Agricultural Extension* (pp. 115-134). FAO, 1990.

Hildebrand, P. E. (Ed.). (1986). *Perspectives on farming systems research and extension.* Colorado: Boulder. Lynne Rienner.

Howell, J. (1989). Public Investment in Africa's Extension Services. In Nigel Roberts (Ed.). *Agricultural Extension in Africa–A World Bank Symposium* (pp. 85-91). Washington DC: The World Bank.

India, Ministry of Agriculture, (1988). *Andhra Pradesh Agricultural Extension Project: An Evaluation Study* [Computer Programme]. CAB: 1R Rural Extension, Education and Training Abstract 1989 012-00365, AN: R 594620.

International Agricultural Development Service (1980). *Agricultural Development Indicators—A Statistical Handbook.* New York.

International Monetary Fund (1989). *International Financial Statistics Yearbook 1989,* Vol. XLI. Washington DC: International Monetary Fund.

Jamison, D. T. and Lau, L. J. (1982). *Farmer education and farm efficiency.* Baltimore: The Johns Hopkins University Press.

Jones, S. & Jones, R. (1985). *Making Agricultural Extension Training Practical: From Evaluation to Curriculum Change in Southern Sudan* [Computer Programme]. CAB: 1R Rural Extension, Education and Training Abstracts 1985 008-006316. AN: R 0087.

Kelsey, L. D. & Hearne, C. C. (1963). *Cooperative Extension Work.* NY: Ithaca.

Land, K. C. (1981). Identification of Socio-Economic Indicators for use in Integrated Social Economic Planning, in UNESCO, *Socio-Economic Indicators for Planning: Methodological Aspects and Selected Samples* (Socio-Economic Studies 2). Paris, France.

Lele, U. (1975). *The design of rural development lessions from Africa.* Baltimore, MD: The Johns Hopkins University Press.

Leonard, D. K. (1977). *Reaching the Peasant Farmer-Organization Theory and Practice in Kenya.* Chicago: University of Chicago Press.

Luning, H. A. (1984). Training qualified agricultural and rural development staff in low-income countries—Some experiences from international training programmes. *Training for Agricultural and Rural Development 1983* (FAO Economic and Social Development Series No. 31). Rome: Italy.

Maalouf, W. D. (1989). The FAO's Experience in Agricultural Extension for Agricultural and Rural Development. In B. E. Swanson (Ed.). *Report of the Global Consultation on Agricultural Extension* (pp. 171-204). FAO, Rome, 1990.

Marshall, A. (1920). *Principals of Economics*, 8th ed. New York: Macmillan.

McCabe, M. S. & Swanson, B. E. (Eds.) (1975). *International Directory of Extension Organizations and Extension Training Institutions*. University of Wisconsin-Madison: Midwestern Universities Consortium for International Activities, Inc.

McKee, C. (1984). Methodological Challenges in Analysing the Household in Farming Systems Research: Intra Household Resource Allocation. In C. Flora (Ed.). *Proceedings of the Kansas State University's 1983 Farming Systems Research Symposium*. Manhattan, Kansas State University.

Morita-Lou, H. (Ed.). (1985). *Science and Technology Indicators for Development*. Boulder, Co: Westview Press.

Mosher, A. T. (1958). *Varieties of Extension Education and Community Development*. Ithaca, NY: Cornell University Comparative Extension Publication, No. 2.

Nyiira, Z. M. (1987). Policy structures and crop productivity in Third World Countries. *Agricultural Administration & Extension 27*, 1-11.

Oram, P., Zapata, J., Alibaruho, G. & Roy, S. (1979). *Investment and Input Requirements for Accelerating Food Production in Low-Income Countries by 1990*. Washington, DC: International Food Policy Research Institute (IFPRI).

Pardey, P. G. & Roseboom, J. (1989). *ISNAR agricultural research indicator series 1989*. International Service for National Agricultural Research: NY: Cambridge.

Peterson, W. E. Sands, C. M. and Swanson, B. E. (1989). *Technology development and transfer systems in agriculture*. The USAID-INTERPAKS Cooperative Agreement Technical Report. International Programme for Agricultural knowledge System. University of Illinois, Urbana-Champaign.

Potsh, B. (1985). Female farmers, mothers-in-law and extension agents: Development planning and a rural LUO community in Kenya. In Reta S. Gallin & Anita Spring (Eds.) *Women creating wealth*: *Transporting economic development*. Association for women in development conference, Washington, DC.

Roberts, N. (Ed.). (1989). *Agricultural extension in Africa—A World Bank Symposium*. Washington, DC: The World Bank.

Roling, N. (1988). *Extension Science—Information Systems in Agricultural Development*. New York: Cambridge.

Rolls, M. J., Jones, G. E., & Garforth, C. (1988). The dimensions of rural extension. *Reading Rural Development Communications Bulletin*, pp. 3-6.

Schultz, T. W. (1956). *Agriculture and the Application of Knowledge. A look to the Future* (Battle Creek, Mich.: W. K. Kellogg Foundation, pp. 54-78).

Schultz, T. W. (1971). *Investment in Human Capital—The Role of Education and Research*. New York: The Free Press.

Schultz, T. W. (1981). *Investing in People—The Economics of Population Quality*. Berkeley: University of California Press.

Spicer, H. E. (1986). Human problems in technological change. In Peter E. Hildebrand (Ed.). *Perspectives on Farming Systems Research and Extension* (pp. 104-117). Colorado: Boulder. Lynne Rienner.

Statistical Package for the Social Sciences Inc. (1988). *SPSS-X Trends*. Illinois: Chicago.

Staudt, K. A. (1975-76). Women farmers and inequalities in agricultural services. *Rural Africa*. 29 (1975-76)

Swaminathan, M. S. (1985). Proposed recommendations for discussion. In the Rockfeller Foundation and International Service for National Agricultural Research. *Women and Agricultural Technology: Relevance for Research*. Volume 1-Analysis and Conclusions (pp. 51-73).

Swanson, B. E. (Ed.). (1984). *Agricultural Extension—A reference manual (2nd Ed.)*. FAO

Swanson, B. E. and Claar, J. B. (1984) The history and development of agricultural extension. In B. E. Swanson (Eds.). *Agricultural Extension: A reference manual (2nd Ed.)* Rome: FAO

Swanson, B. E., Farner, B. J. and Bahal, R. (1990). *International Directory of Agricultural Extension Organizations*. Rome: FAO

Swanson, B. E., Rassi, Jaffar (1981). *International Directory of National Extension Systems*. University of Illinois: Bureau of Research.

Taylor, C. L. (Ed.). (1980). *Indicators systems for political, economic and social analysis*, (pp. 11-21). Cambridge, MA: Oelgeschlager, Gunn, & Hain.

True, A. C. (Ed.) (1980). *A history of agricultural extension work in United States 1785-1923*. US. Department of Agriculture, Miscellaneous publication No. 15, Washington DC: Government Printing Office.

UNESCO (1987). *Statistical Year Book*. Paris: Author.

Van Apeldoorn, G. J. & Turner, S. D. (1984). *Research on the rural poor of Lesotho: Preliminary indicators and future directions* (Computer Programme). CAB: Abstract, NN: R306471.

Van den Ban, A. W. & Hawkins, H. S. (1988). *Agricultural Extension*. New York: Longman Scientific & Technical.

Vengroff, R. (1984). *The Administration of Rural Development: The Role of Extension Agents in Upper Volta and Zaire* [Computer Programme]. CAB: 1R Rural Extension, Education and Training Abstracts 1984 007-01042; 2R, AN: R160205.

von Blanckenburg, P. (1982). Basic concepts of agricultural extension in developing countries. *Agricultural Administration* 10(1), 35-43.

von Blanckenburg, P. (1984). *Agricultural extension system in some African and Asian countries—An analysis of country reports.* Economic and Social Development paper 46. FAO

von Blanckenburg, P. (1988). *Agricultural Extension systems in some African and Asian countries—An analysis of country reports.* Economic and Social Development paper 46. FAO

Whyte, W. F. (1986). The need for a new strategy. In Peter E. Hildebrand (Ed.). *Perspectives on Farming Systems Research and Extension* (pp. 1-12). Colorado: Boulder. Lynne Rienner.

World Bank (1982). *Women in development.* Washington, DC. World Bank.

World Bank (1982). *World Development Report 1982.* NY. Oxford University Press.

World Bank (1985). *Agricultural Research and Extension–An evaluation of the World Bank's experience.* USA: Washington, DC, the World Bank.

World Bank (1989). *World Development Report 1989.* NY: Oxford University Press.

World Bank (1990). *World Tables 1990-91 Edition.* Baltimore: The Johns Hopkins University Press.

INDEX